Waste-to-Energy

Lisa Branchini

Waste-to-Energy

Advanced Cycles and New Design Concepts for Efficient Power Plants

Springer

Lisa Branchini
CIRI- Energy and Environment
University of Bologna
Bologna
Italy

ISBN 978-3-319-13607-3 ISBN 978-3-319-13608-0 (e-Book)
DOI 10.1007/978-3-319-13608-0

Library of Congress Control Number: 2015933516

Springer Cham Heidelberg New York Dordrecht London

Printed on acid-free paper

Springer is part of Springer Science+Business Media (www.springer.com)

To my parents,
who always gave me the chance to choose.
"Nothing in life is to be feared, it is only to be understood."

Marie Sklodowska Curie

Preface

Who Should Read This Text?

This text is intended for students and engineers interested in the design of classical and novel waste-to-energy (WTE) systems. It focuses on: (i) the description of processes involved in the conversion of municipal solid waste (MSW) into energy; (ii) key problems concerning corrosion, which negatively affect WTE steam cycle design parameters; (iii) possibilities to increase WTE conversion efficiency and performance through innovative cycles based on integration with gas turbines (GT).

This text has evolved from 3 years of PhD research studies carried out at the DIN—Department of Industrial Engineering, Alma Mater Studiorum—Università di Bologna.

The target readers are graduate students, masters students, or PhD students with a thorough understanding of basic thermodynamic processes in energy systems, and engineers working in the field of energy systems and MSW conversion. Common prerequisites are knowledge of general mechanical engineering topics such as thermodynamics, fluid machines, energy systems, etc.

Why Has This Text Been Written?

Several countries have waste management systems with significant improvement boundaries to reach worldwide state-of-the-art engineering, health, and safety standards. An increase in environmental and health concerns, combined with new possibilities to exploit these ordinary life by-products as a valuable energy resource, has led to the exploration of alternative methods for waste disposal. In this context, the energy conversion of MSW in WTE power plants is one of the principal means of integrated waste management; its potential is increasing throughout Europe, both in terms of plant number and capacity, furthered by legislative directives. The aim of the new Waste Framework Directive 2008/98/EC is to drive WTE facilities to maximize energy recovery and utilization of waste heat, in order to substitute energy produced with conventional fossil fuel-fired power plants. This calls for novel approaches to maximize the conversion of MSW into energy. In particular, the idea of an integrated configuration made up of a WTE and a GT derives from the desire to eliminate, or at least mitigate, limitations affecting the WTE conversion processes bounding the thermodynamic efficiency of the cycle. The aim of this book is to study and propose innovative integrated WTE cycles. It investigates, from a thermodynamic point of view, integrated WTE–GT system sharing, particularly the steam cycle. The book also proposes and analyzes novel approaches to maximize the conversion of MSW into energy, and discusses, in its final chapters, the difficult task of optimizing conversion efficiency for multi-fuel (MF) energy systems.

What Can Be Learned from This Text?

The primary emphasis explains the basic concepts related to WTE conversion processes, highlighting the most relevant aspects limiting the thermodynamic efficiency of a WTE power plant. In particular, a preliminary thermodynamic analysis is carried out to quantify the influence of the main steam cycle parameters and plant configurations for WTE conversion efficiency. The book's core lesson describes and discusses the integrated WTE–GT concept. Two basic types of hybrid dual-fuel combined cycle arrangements are explained: (i) *steam/waterside integration* and (ii) *windbox repowering*. Focusing on steam/waterside integration, various layouts are proposed; for each analyzed configuration, the optimum WTE–GT plant to maximize steam generation and power output, and to minimize discharged outlet temperatures is identified.

Issues of power output allocation and conversion efficiency for an MF energy system, receiving different type of input fuels and generating a single useful output, are discussed. Based on several performance indexes, specifically introduced to evaluate an MF energy system, different approaches are compared in order to evaluate the WTE–GT integrated system performance.

Finally, a specific case study of a medium-sized WTE facility integrated with different market-available GT units is presented; for each considered, GT units and integrated plant configurations, as well as detailed modifications to the WTE original steam cycle and the resultant achieved performance gains are discussed.

Acknowledgments

A number of people have directly or indirectly contributed to this text. I would like to thank, in particular, my advisors, Prof. Michele Bianchi and Prof. Antonio Peretto, for pushing me to keep learning. Sincerest thanks are due to my colleagues, Francesco and Andrea, for their ongoing support, and advice since I was a student. I would like to thank Springer, the publisher of the book, and my editors, Tiffany

Gasbarrini and Rebecca Hytowitz for giving me this opportunity and guiding me through every step of the way from the beginning. Finally, a special thanks goes to my friend and colleague Elisa for her continuous support and friendship over all these years.

Bologna, October 2014 *Lisa Branchini*

Book organization

The manuscript is divided into four main parts.

Part I: Presents the outline of the scenarios of interest and describes the fundamental key aspects concerning the waste-to-energy conversion processes. Chap. 2 presents an overview of municipal solid waste production and disposal for both the European and the Italian markets. Chapter 3 introduces and describes the basic concepts related to waste-to-energy conversion processes, highlighting the most relevant aspects circumscribing the thermodynamic efficiency of a WTE power plant.

Part II: In Chap. 4, a thermodynamic analysis is carried out in order to quantify the influence of steam cycle parameters and plant configurations on WTE thermodynamic efficiency. The aim of this preliminary analysis is to understand and compare possibilities and benefits of a thermodynamic cycle upgrade for a WTE facility.

Part III: Explains and discusses the hybrid combined cycle (HCC) concept. Two basic types of hybrid dual-fuel combined cycle arrangements are detailed in Chap. 5: (i) *steam/waterside-integrated* HCC and (ii) *windbox repowering.*

Steam/waterside integration represents a cost-effective method to increase the electric efficiency of a WTE power plant without the necessity of new expensive materials to prevent corrosion risks. Indeed, a further superheating of the steam is performed in a separated heat recovery steam generator (HRSG), by means of a high-grade fuel, that does not cause corrosion problems in the superheater banks.

Windbox repowering, however, due to a considerable increase in combustion gas volumetric flow rate, requires a significant redesign of combustion, convective, and gas cleaning sections.

With reference to WTE–GT steam/waterside integration, the logic governing plant match in terms of steam production as the function of thermal power is investigated and explained. The thermodynamic analysis carried out in Chap. 6 assesses and defines, for given layout and operative conditions, the optimum WTE–GT plant match to maximize steam generation and power output, and to minimize discharged outlet temperatures. Moreover, several proposed WTE–GT steam/waterside-integrated layouts, with reference to one pressure level heat recovery boiler configuration, are presented and detailed.

Part IV: In Chap. 7, open research issues to evaluate the performance and the efficiency conversion of multi-fuel (MF) energy systems are discussed. The difficulty in defining a performance index capable of quantifying the efficiency of the integrated system, compared to separate generation, lies in assigning the extra power generated as a consequence of systems integration. Thus, in Chap. 7, several performance indexes are proposed and described. Finally, in Chap. 8, detailed modifications to the WTE cycle and the resulting enhancement of its performance are presented for two different WTE–GT integrated cases. Numerical results for a representative WTE plant integrated with different commercial GT units are shown and discussed. The performance indexes and output allocation approaches defined in Chap. 7 are applied to quantify, for each case, integrated system performance.

Contents

List of Abbreviations

BC	Bottomer cycle
C	Compressor
CC	Combined cycle
CHP	Combined heat and power
DEA	Deaerator
ECO	Economizer
EVA	Evaporator
EU	European Union
FGR	Flue gas recirculation
GT	Gas turbine
HCC	Hybrid combined cycle
HE	general heat exchanger
HP	High pressure
HRSG	Heat recovery steam generator
IBC	Inverted Brayton cycle
ICE	Internal combustion engine
IP	Intermediate pressure
LHV	Lower heating value
LP	Low pressure
MF	Multi-fuel
MSW	Municipal solid waste
ORC	Organic Rankine cycle
PL	Pressure level
RH	Reheater
SF	Single fuel
SH	Superheater
ST	Steam turbine
T	Turbine
TC	Topping cycle
WFD	Waste framework directive
WTE	Waste-to-energy

Symbols

c	specific heat [kJ/kgK]
E	electric energy output [MWh]
F	fuel input power [MW]
f_b	equivalent factor accounting for energy losses [–]
h	specific enthalpy [kJ/kg]
I_{imp}	Imported Enery [MWh]
m	mass flow rate [kg/s]
p	pressure [bar]
P	power output [MW]
Q	heat [MW]
SI	Synergy Index [–]
T	temperature [°C]
U	total useful electric energy output [MWh]

Greek Symbols

ε	electric equivalent conversion factor [–]
η	efficiency [–]
Δ	Difference
ξ	HRSG effectiveness [–]
x	fuel input weighting factor [–]

Subscripts and Superscripts

I	first law
boil	Waste-to-Energy boiler
C	Carnot
EE	Electric Equivalent
EXH	exhausted
ev	evaporation
i	generic i-th input
in	input
it	isentropic efficiency
j	generic j-th input
lat	latent heat
max	maximum
nu	non useful heat

O	Outlet
Prod	produced
rec	recoverable
ref	reference
s	steam
sc	sub-cooling
TOT	Total
W	Waste

Part I
WTE State-of-the-Art

Chapter 1
Introduction

In recent years, the increased focus on energy resources and environment has changed perceptions about waste. The growth in economic activities, standard of living, and population has led to a sharp increase in the quantity of generated waste. Waste poses a highly complex and heterogeneous environmental problem since all human activities, inevitably, result in generation of waste due to imperfect utilization of energy and resources.

Currently, refuse disposal methods are determined by cost and their effect on the environment. Intensification of environmental and health concerns, combined with the possibility of exploiting day-to-day by-products as valuable energy resources, has led to the exploration of alternative methods for waste disposal. Energy conversion of municipal solid waste (MSW) in waste-to-energy (WTE) power plants is one of the principal means to integrated waste management; its potential is increasing throughout Europe, both in terms of plant number and capacity and is furthered by legislative directives.

Waste Framework Directive (WFD) 2008/98/EC [1] sets the basic concepts and definitions related to waste management and lays down waste management principles, introducing the so-called "waste hierarchy" in which waste prevention is still considered the main goal followed by reuse and recycle. Nevertheless, growing interest in the WTE conversion process is evident.

WTE plants have been established worldwide as a sustainable method for the disposal of residual waste.

Big investments were made in new WTE facilities and recycling activities in many European Union (EU) countries at the beginning of this century. There are enough incineration and recycling capacities in Central Europe to prevent landfilling of organic waste. Since huge amounts of greenhouse gas (GHG) emissions have been prevented by WTE facilities as compared to landfills, the unused capacities of WTE installations should be used to recover energy from waste in EU Member States where waste disposal is predominantly through landfills.

L. Branchini, *Waste-to-Energy*, DOI 10.1007/978-3-319-13608-0_1

The possibility of generating energy in the form of heat, electricity, or combined heat and power (CHP) became an attractive way of dealing with waste. Although various WTE technologies such as waste incineration, anaerobic digestion, gasification, etc. exist, mass incineration is still the most common.

Incineration plants, which were a viable option earlier, are the predecessors of the present day's WTE plant. Heat recovery mechanisms were added to the incinerators in the form of waste heat boilers and later in the form of low pressure and temperature steam boilers. These incinerators then evolved into waterwall boilers with an integrated furnace.

At present, direct combustion over a moving grate with the generation of superheated steam feeding a steam turbine in a *Hirn* cycle is the dominant technology used to recover energy from MSW. The amount of energy recovered from the MSW combustion can vary significantly with the characteristics of the MSW fed into the boiler (composition, mass flow rate, and heating value), the combustion technology, the configuration and features of the recovery boiler (adiabatic or integrated), and the characteristics of the thermodynamic cycle. Owing to the heterogeneous nature of waste, there are some differences between a conventional fossil fuel power plant and a WTE plant with respect to the chemical-to-electrical energy conversion process. The thermodynamic efficiency of a WTE power plant is constrained, mainly, by the following aspects [2]:

- Maximum temperature of the steam cycle is limited by corrosion problems, mainly affecting the high-temperature sections.
- The steam-superheated temperature makes it necessary to keep the evaporative pressure low to avoid build-up of liquid at the steam turbine outlet.
- Typically modest power output and steam mass flow rate of a WTE power plant imply low steam turbine isentropic efficiency.

Consequently, the thermodynamic efficiency of WTE power plants typically ranges from 25 to 30 %.

The new WFD 2008/98/EC lays down the criteria for producing energy efficiently from waste (the so-called "R1 formula") to evaluate recovery status of the plant. Although the energy efficiency criteria can only be regarded as a starting point for all the Member States, they are a benchmark for quality standards in waste recovery. The energy recovery status provides incentives for future investments in WTE plants, especially those located close to customers. The aim of the Directive is to spur WTE facilities to maximize energy recovery and waste heat utilization so they can eventually substitute conventional fossil fuel-fired power plants.

There is substantial scope for improvement in WTE systems. This calls for novel approaches to maximize the conversion of MSW into energy.

This text investigates, from a thermodynamic point of view, the scope for performance improvement in WTE plants, starting with the use of state-of-the-art technologies. Both repowering and advanced WTE options are analyzed.

In particular, the idea of an integrated configuration made up of a WTE and a gas turbine (GT) derives from the desire to eliminate, or at least mitigate, the limitations of the WTE conversion processes bounding the thermodynamic efficiency of the cycle.

The study on the integrated WTE–GT system, that is, sharing (i) the steam cycle, (ii) the flue gas paths, or (iii) combining both the mechanisms, is the book's core. The thermodynamic analysis investigates and defines the logic governing plants and it should match in terms of steam production and steam turbine power output as a function of the thermal power generated. The book presents detailed discussions on the modifications to a typical WTE steam cycle arrangement in order to evaluate the resulting performance enhancement. Numerical results of a conventional reference WTE plant integrated with different GT commercial units are discussed.

This book aims to provide guidelines for designing advanced WTE solutions for future implementation in new and/or existing plants of industrial partners and other similar industrial arrangements.

Another purpose of the book is to discuss meaningful performance indicators of systems such as multi-fuel (MF) energy systems. Performance indexes, specifically introduced to assess the performance of the proposed integrated configuration and to determine the power output of each input fuel, have been presented and applied to the plant under consideration.

The framework of this research integrates a collaborative project involving *Alma Mater Studiorum—Università di Bologna—DIN* and a leading Italian multiutility company operating, particularly, in the field of design and management of WTE facilities. This joint research project was aimed at assessing the possibility of integrating new and existing WTE power plants with GT units.

References

1. Directive 2008/98/EC of the European Parliament and of the Council of 19 November 2008 on waste. Official Journal of the European Union OJ L 312/3, 22.11.2008
2. Bianchi M, Branchini L, De Pascale A, Falchetti M, Fiore P (2014) Advanced waste-to-energy steam cycles. Energy Procedia 45:1205–1214. doi:10.1016/j.egypro.2014.01.12

Chapter 2
Municipal Waste Overview

Improved waste management is an essential component in order to make countries more resource efficient. This important target is driving European environmental policies and legislations towards a more efficient waste management system.

In this chapter, the "waste management hierarchy," proposed by the 2008/98/EC European Union Directive, is described. Moreover, in order to provide a general overview on municipal solid waste (MSW), this chapter focuses on waste generation and disposal for both the European and the Italian markets. Information and data concerning European Union (EU) countries' progress in enhancing recycling rates will be discussed, with a focus on on related policies such as landfill average costs and taxes.

2.1 Municipal Solid Waste Definition and Management System Hierarchy

The definition of "municipal waste" used all around the EU countries varies, reflecting different waste management operations. In order to collect and compare yearly reporting data of waste generation and management from European countries, "municipal waste" is defined as follows [1]: "Municipal waste is mainly produced by households, though similar wastes from sources such as commerce, offices and public institutions are included. The amount of municipal waste generated consists of waste collected by or on behalf of municipal authorities and disposed of through the waste management system."

In this context, municipal waste is understood as waste collected by or on behalf of municipalities.

In the EU's Landfill Directive, MSW is defined as [2] "waste from households, as well as other waste which, because of its nature or composition, is similar to waste from households."

L. Branchini, *Waste-to-Energy*, DOI 10.1007/978-3-319-13608-0_2

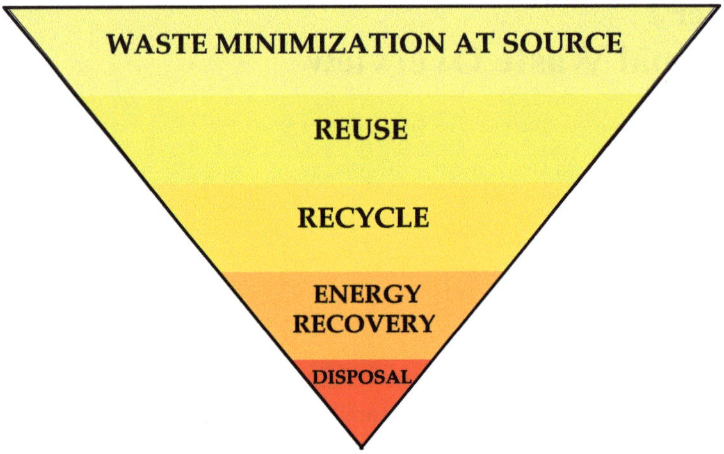

Fig. 2.1 Waste management hierarchy as suggested by 2008/98/EC Directive

Thus, the definition refers to "household waste and similar waste." The typical MSW composition contains food and kitchen waste, green waste, glass, paper, plastics, as well as other recyclable materials, etc.

The revised 2008/98/EC Directive also sets the basic concepts and definitions related to waste management and lays down waste management principles such as the "waste hierarchy" [3]. The EU's approach to waste management is based on three fundamental principles: waste prevention, recycling and reuse, and improving final disposal and monitoring. Based on the EU's approach to waste management, the best and most economical way of dealing with waste is to minimize its production; if waste cannot be prevented, as many of the materials as possible should be recovered, preferably by recycling. Where possible, waste that cannot be recycled or reused should be safely incinerated, recovering the energy released with waste combustion and leaving landfill as the last option for waste disposal.

Figure 2.1 schematically shows, through an inverted pyramid, the waste management hierarchy suggested by European Commission's directives.

European environmental policies and national efforts must be in the direction of shifting waste management up the waste hierarchy—reducing waste disposal (i.e., landfilling) while focusing on waste prevention, reuse, recycling, and energy recovery.

Development and progress on waste management for each one of the 32 European Environment Agency (EEA) member countries[1] will be discussed in the next paragraphs, reviewing the latest available data.

[1] The EU-27, Iceland, Norway, Switzerland, Turkey, and Croatia.

2.2 Overview of Waste Production and Disposal for European Countries

The socioeconomic growth of the most industrialized countries has involved a progressive increase of MSW production. In the EU-27, the average MSW production was about half a metric ton per person in 2010 [4]; hence, issues related to the disposal of MSW turn out to be very important. Figure 2.2 shows the amount of MSW produced per person for the 32 EEA members in 2001 and 2010.

A comparison of yearly MSW generation data shows that most of the countries (21 of 32) generated more municipal waste per capita in 2010 than in 2001, while 11 reduced municipal waste per capita.

Despite the problem that the MSW definition differs between countries (which means that data should be used with caution), a very wide range of waste generation can be observed.

The EU-27 average value equaled 502 kg/p.p. in 2010; countries such as Cyprus or Switzerland considerably exceed the EU average value, whereas Eastern

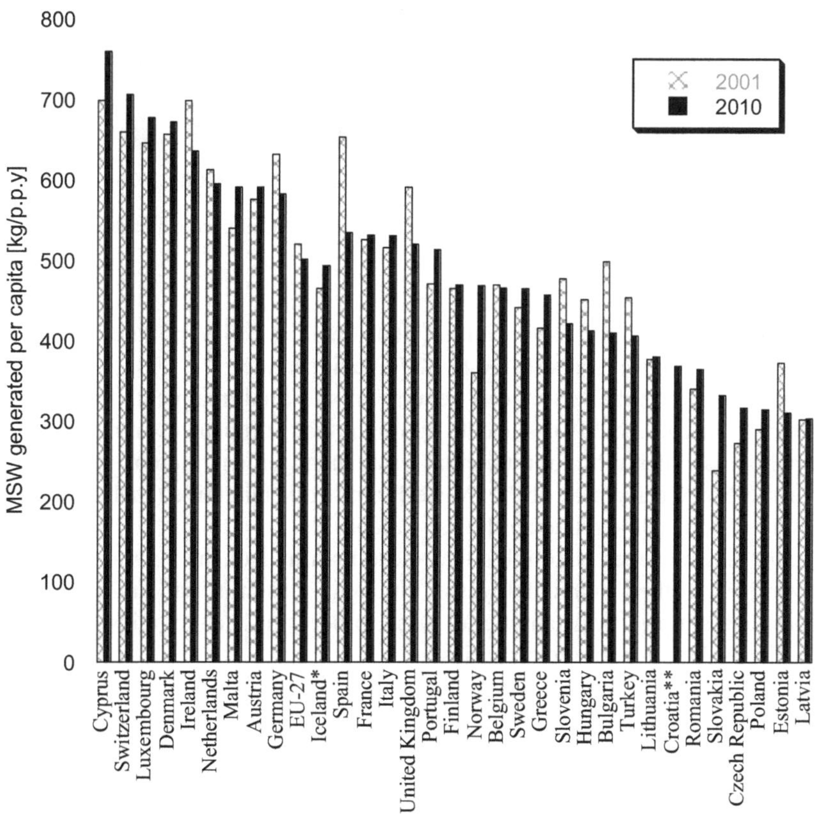

Fig. 2.2 Municipal waste generated per capita in 2001 and 2010 [4]. (Iceland: 2008 data used for 2010; Croatia: 2004 data used for 2001)

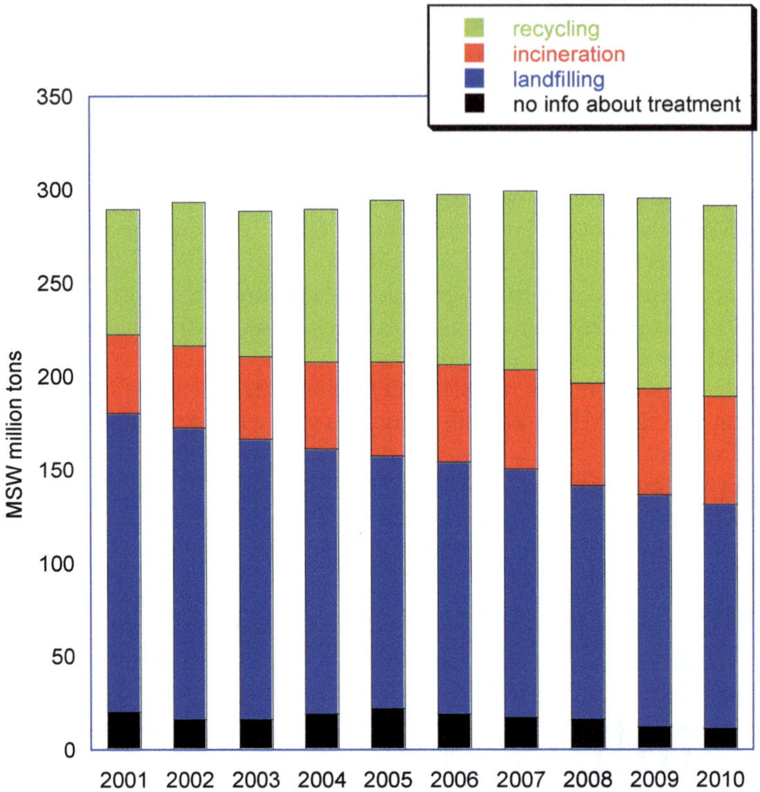

Fig. 2.3 Development of municipal waste management in 32 European countries, from 2001 to 2010 [4]

European countries have lower values. Italy, with a MSW production of about 530 kg/p.p./year, lies just above the EU-27 average value.

Figure 2.3 shows the MSW total generation and final treatments for the 32 EU countries from 2001 to 2010. The data highlight that European countries are climbing up the waste hierarchy for municipal waste management, thereby implementing one or more key principles of the Waste Framework Directive (such as recycling and incineration) while reducing the use of landfills.

Figure 2.3, however, highlights that more than half of the countries still landfill over 50 % of their municipal waste. The main reason is that landfill is still the cheapest and simplest way, among all waste-processing technologies, to deal with waste disposal in most of the countries.

In Fig. 2.4, percentages of MSW final treatments in the EU-27 countries are reported in detail. Only few exceptions (such as the Netherlands, Denmark, and Sweden) have a high level of alternatives for final treatment disposal. The percentage of incinerated waste ranges from zero (e.g., East European countries) to over 50 % (e.g., Denmark) with an average value equal to about 18 % for 27 EU countries in

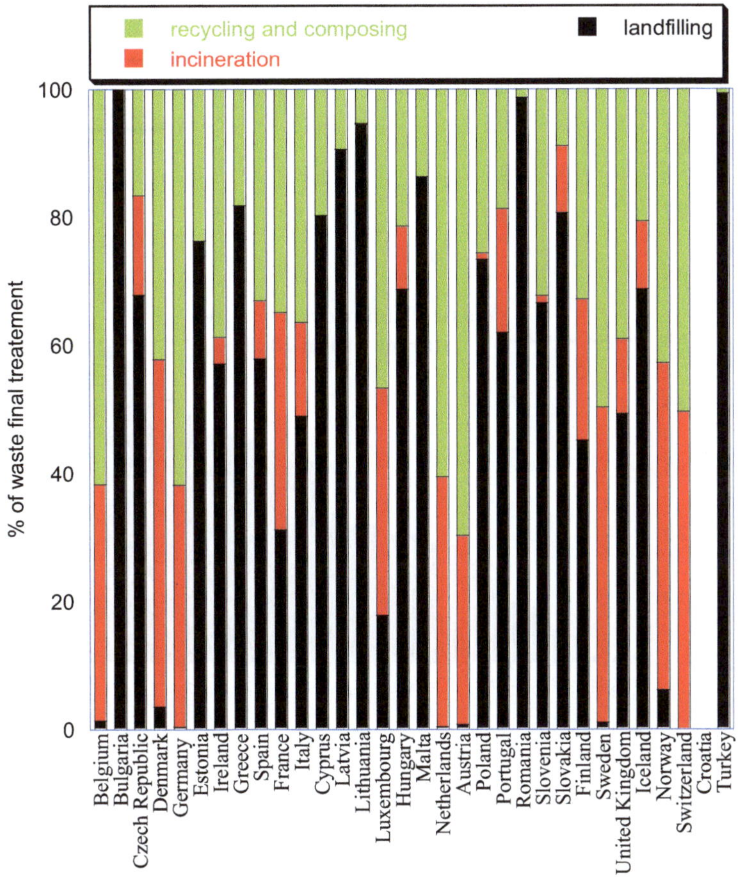

Fig. 2.4 Final treatment percentage allocation for 27 European countries in 2010 [4]. (No data is available for Croatia)

2010. In general, during the period from 2001 to 2010, the increase in recycling and incineration of MSW has led to a decrease in landfill waste (see Fig. 2.3).

Figure 2.5 compares MSW recycling rates (including also composting and digestion of organic waste) for EEA countries in 2001 and 2010. In the analyzed period, although percentage rates show considerable differences in performance between nations, 22 countries have increased their recycling performance between five and ten percentage points. This clearly indicates a significant improvement in recycling performance. Although, five countries (Switzerland, the Netherlands, Belgium, Austria, and Germany) have already achieved the 50 % recycling target imposed by the European 20-20-20 targets, the majority of countries will need to make an extraordinary effort in order to achieve this goal.

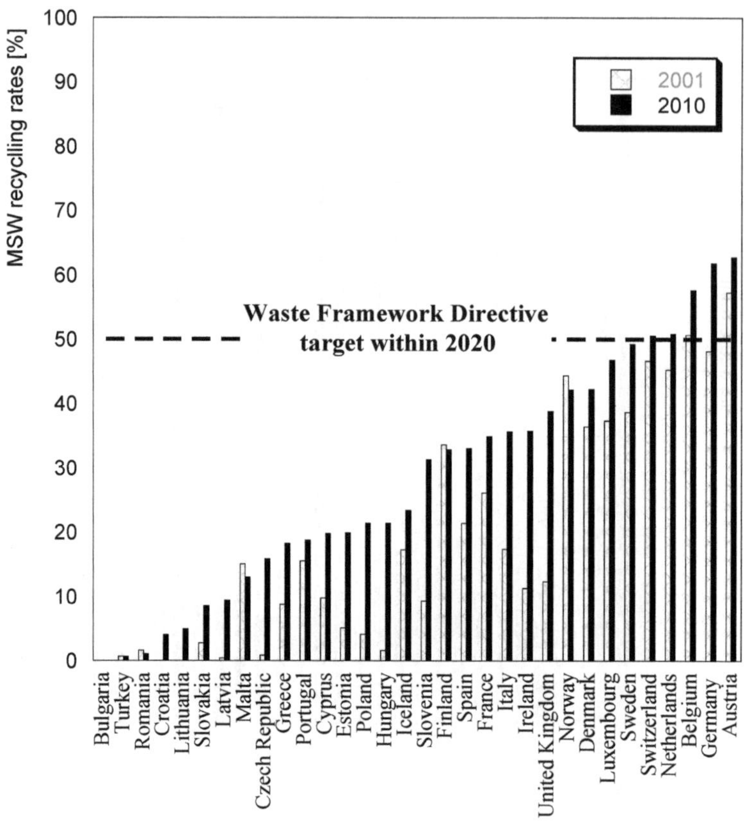

Fig. 2.5 MSW recycled for 32 European countries in 2001 and 2010 [4]. *WFD* Waste Framework Directive

2.2.1 Overview of Municipal Solid Waste Production and Disposal in Italy

Figures 2.6 and 2.7 present the data of MSW total generation from 1998 to 2013 [5] and of MSW generated per capita in 2009 and 2013, focusing on the Italian national and regional markets. According to the report of Istituto Superiore per la Protezione e la Ricerca Ambientale (ISPRA), there are remarkable differences in per capita production across regions. In 2009, central regions had the highest per capita production of waste, equal to about 604 kg/p.p.y, whereas lower values were recorded for southern regions, namely, 493 kg/p.p.y. Emilia Romagna (with 666 kg/p.p.y) had the highest MSW production per habitant, followed by Toscana (with 663 kg/p.p.y). In 2013, the total generation in Italy was about 29,595 metric tons of municipal waste.

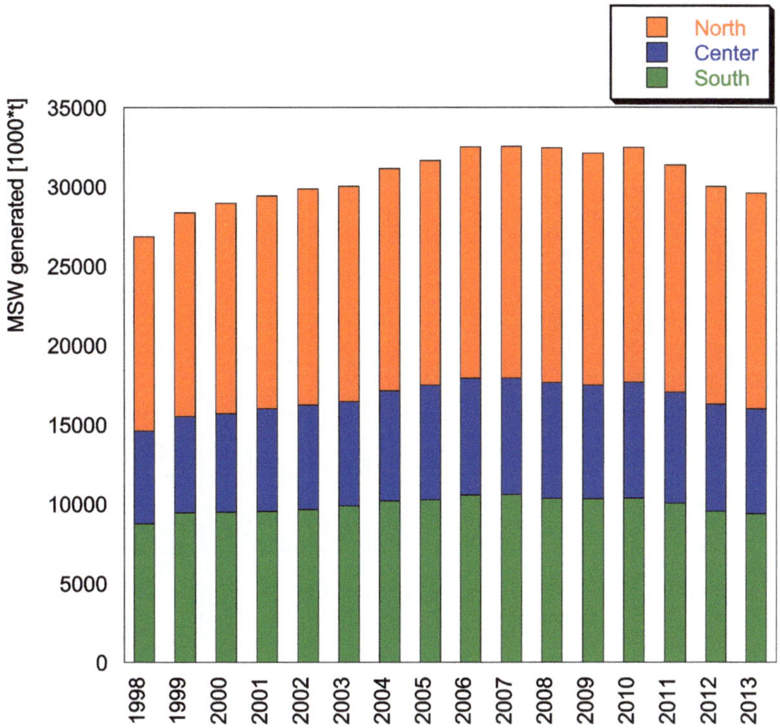

Fig. 2.6 MSW produced in Italy from 1998 to 2013 [5]

In 2009, landfills received around 45 % of the total MSW managed, as shown in Fig. 2.8. Comparing the regional data for waste disposal in 2009, Fig. 2.8 highlights an inhomogeneous situation between regions—northern regions show reduced use of landfill (Lombardia records the lowest use representing 7 % of its total production). An exception, in the southern regions, is represented by Sardegna where regional legislation is driving towards recycle and reuse of waste.

2.3 Average Costs of Municipal Solid Waste Landfill

Cost of various waste disposal methods varies depending on the technology adopted and on the country's specific policy measures (such as taxes). As a rule of thumb, incineration costs are twice the costs of landfill. A recent study [6] quantifies incineration costs between 100 and 250 €/t of waste, whereas costs for landfill have a range down to 20 €/t of waste. Thus, the main reason why landfill is still the predominant treatment option for most of the EU countries is primarily economic. Taxes and charges tend to drive the economics of recycling and composting.

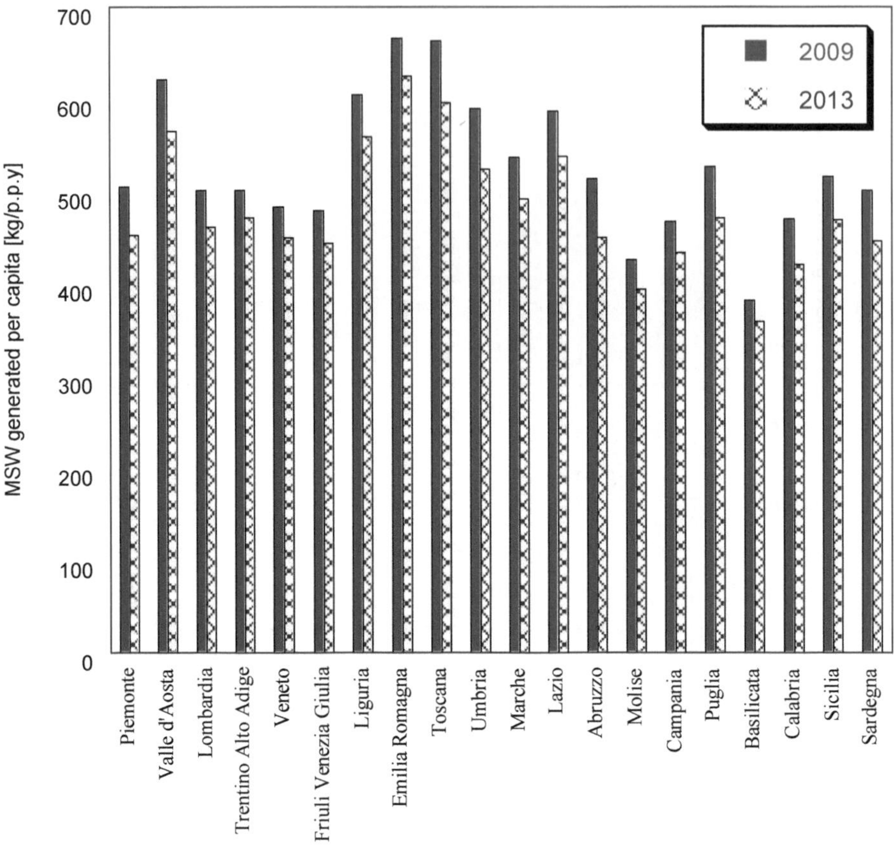

Fig. 2.7 MSW produced per capita in Italian regions in 2009 and 2013 [5]

The most relevant case is Denmark, where cost for landfilling both household waste and hazardous[2] waste drastically increased in the past 17 years [7]. Table 2.1 collects available cost data for some of the EU countries.

A correlation between recycling rates (see Fig. 2.5) and landfill costs is evident. Nations which have sharp policy measures, such as additional landfill taxes or economic support to build up recycling infrastructures, are expected to achieve high recycling rates. As shown in Table 2.1, there is a general trend in increasing costs for waste disposal throughout almost all countries considered. Figure 2.9 shows gate fee and landfill taxes for MSW landfilling in EU Member States. EU countries appear much more likely to meet a 50 % recycling target once landfill charges (or the cost of the cheapest disposal option) approach EUR 100 per metric ton [8].

[2] Waste derived from common manufacturing and industrial processes, from specific industries and from commercial chemical products, etc. are classified as hazardous.

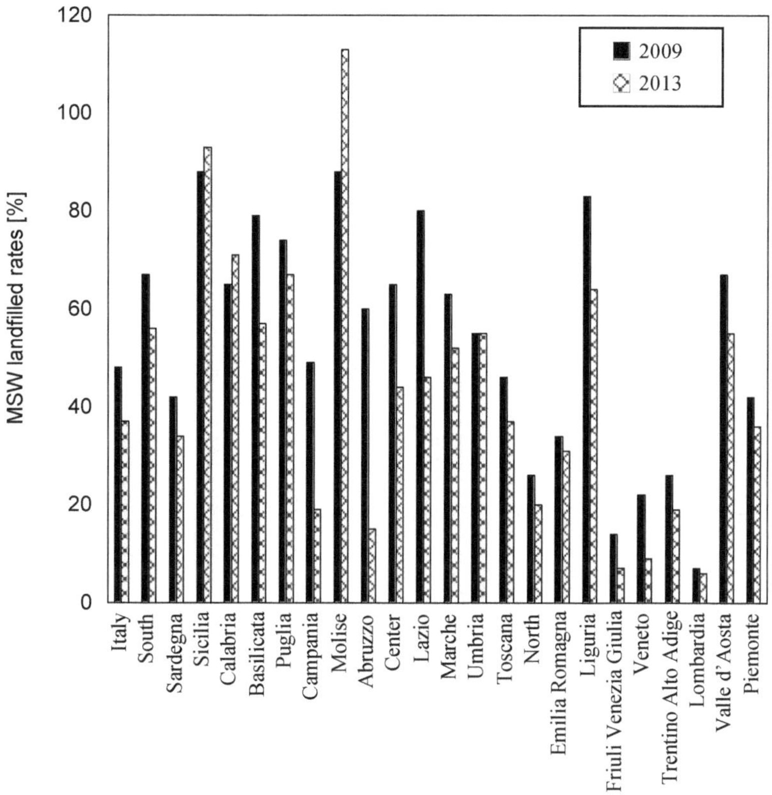

Fig. 2.8 Percentage of landfill disposal of MSW generation in Italian regions in 2009 and 2013 [5]

Table 2.1 Estimated cost range for landfilling of waste (excluding any landfill tax, if relevant) [7]

Country	Latest data	Cost range for MSW [€/t]	% change
Austria	1999	50–150	–
Belgium-Flemish	2003	116	+5 (in 2 years)
Denmark	2004	110	+57 to+340 (in 17 years)
Finland	2003	30–121	–
Germany	2005	123 (average from 12 sites)	–
Greece	2005	8–35	+75 (in 6 years)
Ireland	2005	120–240	+52 (in 4 years)
Italy	2003	90–110	–
Luxembourg	2003	50	–
Netherlands	2002	58	−8 (in 4 years)
Portugal	2004	26	–
Spain	2004	12	–
Sweden	2004	70–90	–
UK	2003	21	–

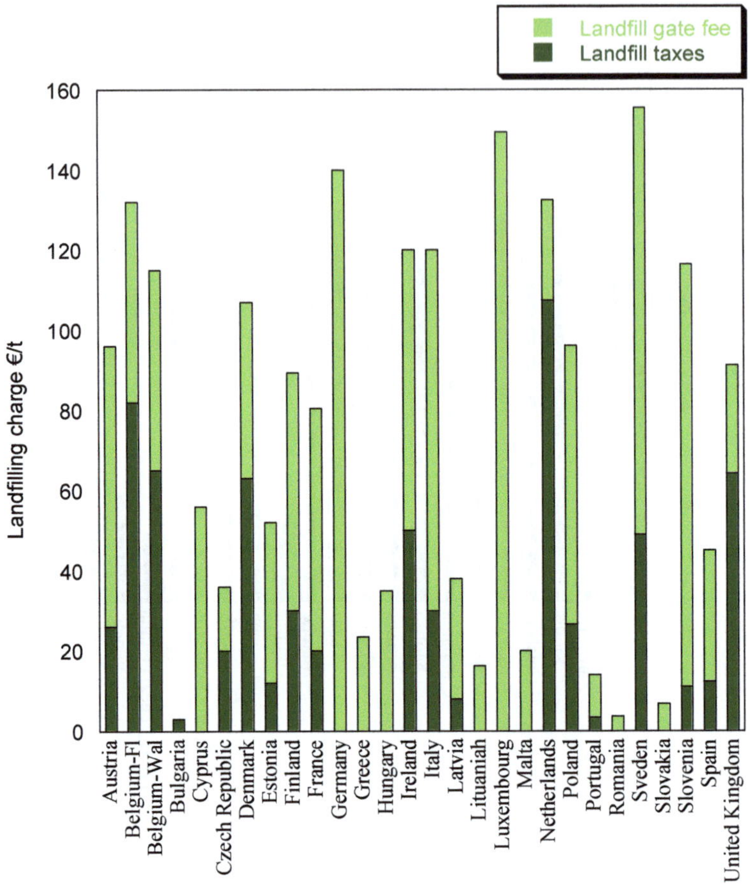

Fig. 2.9 Typical charges (gate fee and taxes) for MSW landfilling in EU Member States and regions [4, 8]

References

1. Eurostat (2012) Reference Metadata in Euro SDMX Metadata Structure (ESMS). http://epp.eurostat.ec.europa.eu/cache/ITY_SDDS/EN/env_wasmun_esms.htm. Accessed 11 Jan 2013
2. EU (1999) Council Directive 1999/31/EC of 26 April 1999 on the landfill of waste, Official Journal of the European Union OJ L 182, 16.7.1999, pp 1–19
3. Directive 2008/98/EC of the European Parliament and of the Council of 19 November 2008 on waste. Official Journal of the European Union OJ L 312/3, 22.11.2008
4. European Environment Agency EEA, Report No 2/2013, Managing municipal solid waste—a review of achievements in 32 European countries. http://www.eea.europa.eu/data-and-maps/figures/municipal-waste-generated-per-capita. Accessed 1 April 2014
5. ISPRA, Istituto Superiore per la Protezione e la Ricerca Ambientale, Edizione (2014) Rapporto Rifiuti Urbani, ISBN: 978-88-448-0665-1. http://www.isprambiente.gov.it/files/pubblicazioni/rapporti/RapportoRifiutiUrbani2014_web.pdf. Accessed 19 Sept 2014

6. Göerner K (2003) Waste Incineration European State of the Art and New Developments. IFRF Combust J Article Number: 200303. ISSN 1562-479X
7. Report on Implementation of the Landfill Directive in the 15 Member States of the European Union; October 2005, European Commission. http://ec.europa.eu/environment/waste/pdf/report_a2.pdf. Accessed 7 March 20134
8. Bio Intelligence Service (2012) Use of economic instruments and waste management performances, Contract ENV.G.4/FRA/2008/0112, European Commission (DG ENV). http://ec.europa.eu/environment/waste/use.htm. Accessed March 2014

Chapter 3
Waste-to-Energy

This chapter introduces and describes the basic concepts related to the waste-to-energy (WTE) conversion processes, highlighting the most relevant aspects that limit the thermodynamic efficiency of a WTE power plant.

An overview of the distribution of WTE plants around Europe, electrical and thermal energy generated, and average plant capacity is provided at the beginning of the chapter. Basic sections and related operations of a WTE facility are described along with typical operative steam cycle parameters. The chapter's final section is dedicated to the definition of the "energy recovery status" (R1 formula) according to the 2008/98/EC Waste Framework Directive (WFD). The thresholds to achieve the energy recovery status are discussed according to WTE plants' operative data.

3.1 Basics of a WTE Power Plant

Waste incineration became established in many European countries at the end of the nineteenth century as a device to minimize waste's volume and to make it hygienic. Since then, as environmental awareness has increased, WTE technology has continuously developed and improved, increasing the importance of efficient energy generation. WTE power plants, in many European countries, have become important tools to generate heat and electricity, replacing the energy produced by conventional fossil fuel power plants.

Various WTE technologies, such as waste incineration, anaerobic digestion, and gasification, exist today; despite that mass burn incineration is still the most common.

The dominant technology for energy recovery from municipal solid waste (MSW) is direct combustion over a moving grate with the generation of superheated steam feeding a steam turbine in a *Hirn* cycle. The amount of energy recovered from the MSW combustion can vary significantly with the characteristics of MSW fed into the boiler (composition; mass flow rate; and lower heating value, LHV), the combustion technology, the configuration and features of the recovery boiler (adiabatic or integrated), and the characteristics of the thermodynamic cycle.

L. Branchini, *Waste-to-Energy*, DOI 10.1007/978-3-319-13608-0_3

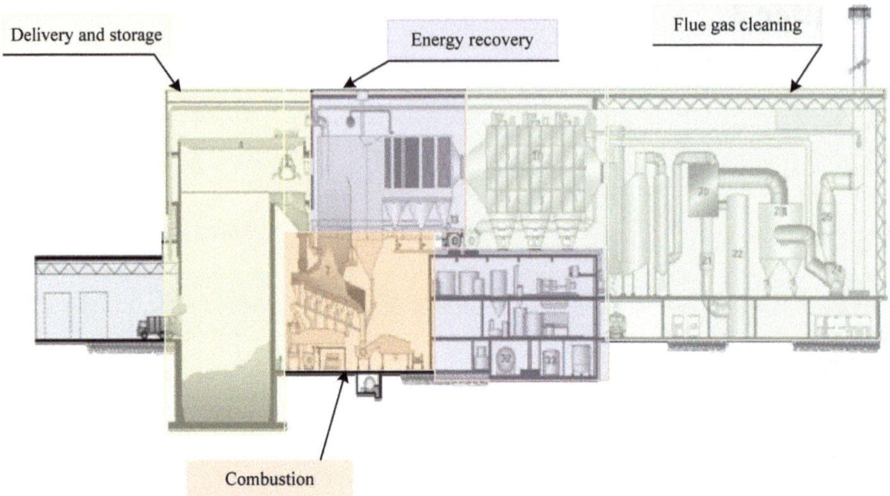

Fig. 3.1 Schematic of a WTE power plant highlighting four macro sections

A simplified and schematic representation of a WTE power plant is shown in Fig. 3.1 highlighting four macro processes and related plant sections:

- Waste delivery and storage section (bunker)
- Waste combustion section (furnace)
- Energy recovery and conversion section
- Flue gas cleaning section

Table 3.1 schematically presents the purpose of the main components of a WTE power plant.

3.1.1 *Waste Delivery and Storage Section*

Trucks, trains, or containers arrive at the delivery area in order to dump the waste into the bunker, usually after visual control and weighing. Enclosure of the delivery area can be an effective means to avoid odor, noise, and emission problems. The bunker is usually a waterproof concrete bed. The waste is piled and mixed in the bunker using cranes equipped with grapples. The mixing of the waste helps achieve a balanced heat value, size, structure, composition, etc., of the material dumped into the combustion section. In order to avoid excessive dust development and gas formation (e.g., methane) from spontaneous chemical reactions between the disposals, as well as the accumulation of odor and dust emissions, waste has to be permanently moved, mixed, and turned by crane operators. Moreover, depending on the calorific value of the waste as well as the layout and the concept of the plant, a fraction of total combustion air (usually the under-grate air) is extracted from the bunker. The

Table 3.1 Purpose of the main components of a waste incineration plant [1]

Objective	Responsibility of
Delivery and storage of waste	Bunker
Mixing of waste	
Feeding of waste to combustion section	
Destruction of organic substances	Furnace
Evaporation of volatile heavy metals and inorganic salts	
Production of potentially exploitable slag	
Volume reduction of residues	
Evaporation of water	Energy conversion section
Recovery of useful energy	
Removal and concentration of volatile heavy metals and inorganic matter into solid residues	Flue gas cleaning
Minimizing emission to all media	

bunker usually has a storage capacity of several days (commonly 3–5 days) of the plan operational throughput; thus, its depth can reach a few dozen meters.

Proper waste feeding from bunker to grate means dosing the right quantity of fuel for steady combustion. Therefore, waste feeding is continuously adjusted as a consequence of achieved combustion temperature and energy production.

3.1.2 The Combustion Section

Basically, waste incineration is the oxidation of the combustible materials contained in the waste. Waste is generally a highly heterogeneous material, consisting essentially of organic substances, minerals, metals, and water. During incineration, flue gases containing most of the available fuel energy as heat are created. Burning of the organic fuel substances occurs once they have reached the necessary ignition temperature and come in contact with oxygen. The combustion process takes place in the gas phase in fractions of seconds and simultaneously releases energy whenever waste calorific value and oxygen supply is sufficient. This can lead to a thermal reaction chain and self-supporting combustion, that is, there is no need to add other fuels.

The main phases of the waste incineration process are [1, 2]:

- *Drying and degassing:* When volatile content is evolved (e.g., hydrocarbons and water) at temperatures generally between 100 and 300 °C. The drying and degassing processes do not require any oxidizing agent and are only dependent on the supplied heat.
- *Pyrolysis and gasification:* Pyrolysis is the further decomposition of organic substances in the absence of an oxidizing agent at approximately 250–700 °C. Gasification of carbonaceous residues is the reaction occurring between water

vapors and CO_2 at temperatures of typically 500–1000 °C. Thus, solid organic matter is transferred to the gaseous phase. In addition to heat, water, steam, and oxygen support this reaction.
- *Oxidation:* When the combustible gases created in the previous stages are oxidized at flue gas temperatures generally between 800 and 1450 °C.

During waste combustion, air is typically added in two main stages: Primary air is generally taken from the waste bunker and supplied through the grate layer into the fuel bed, and secondary combustion air is supplied at high speeds over the grate layer. Additional air (tertiary air) and flue gas recirculation (FGR) can also be used, in particular with the aim to increase turbulence and decrease pollutant emissions.

The primary air is usually preheated either by using combustion grate cooling and/or steam supplied by steam turbine bleed.

Secondary air is blown into the incineration chamber at high speeds via, for example, injection lances. This is carried out to assure complete incineration, and it is also responsible for the intensive mixing of flue gases and the prevention of the free passage of unburned gas streams. Preheating of primary combustion air can have a positive influence on overall energy efficiency in case of electricity production. Moreover, combustion air preheating is particularly beneficial for assisting the combustion of high moisture content waste.

When designing a waste boiler, the design metric tons per day of waste to be combusted and the typical range of heating value that is expected for the refuse in that location need to be known. The boiler is typically designed for the maximum metric ton per day input and the maximum refuse heating value.

Figure 3.2 shows a typical "capacity diagram" for a waste incineration plant where the operating area is highlighted as a function of waste mass flow rate for different waste LHV. The two variables, waste mass flow rate and LHV, identify on the vertical axis the thermal input in the system. From Fig. 3.2, the WTE operating area is defined by:

- *OP1:* Maximum firing rate (100 % capacity) at design point waste input (design LHV, design waste throughput)
- *OP2:* Design LHV waste input at minimum boiler load (e.g., 60 % capacity; design LHV, but minimum waste throughput)
- *OP3:* Maximum boiler load with highest LHV (means reduced waste throughput)
- *OP4:* Minimum boiler load with maximum LHV
- *OP5:* Reduced boiler load with minimum LHV (means maximum waste throughput)
- *OP6:* Reduced boiler load with minimum LHV (means intermediate waste throughput)

Due to the extremely heterogeneous nature of MSW, in contrast to fossil fuel, the operating condition of a WTE plant is continuously adjusted, as much as possible, around the design point (OP1 in Fig. 3.2) in order to maintain nominal plant conditions.

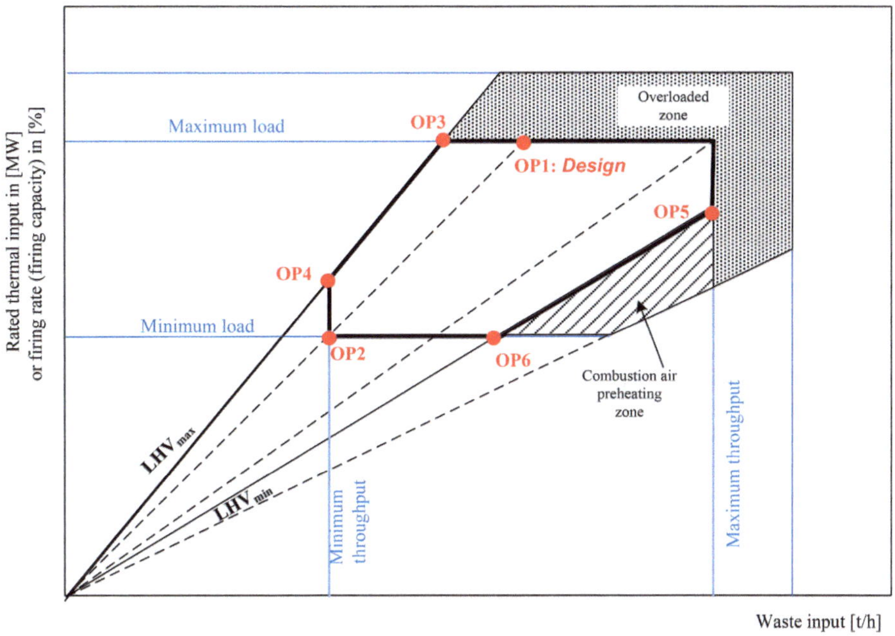

Fig. 3.2 Typical furnace capacity diagram of a waste incineration plant showing the design and off-design operating points

In the normal operating area, two different zones can be detected: a zone where, due to low LVH value, preheating of combustion air is necessary and an overloaded zone where WTE operation can be allowed only for a limited amount of time.

Grate incinerators are widely applied for the incineration of mixed municipal waste. In Europe approximately 90 % of installations treating MSW use grates [1].

Grate incinerators usually have the following components: waste feeder, incineration grate, bottom ash discharger, incineration air duct system, incineration chamber, and auxiliary burners.

Depending on waste LHV, grate cooling can be accomplished with air or a liquid cooling medium (usually water). The flow of the cooling medium is from colder zones to progressively hotter ones in order to maximize the heat transfer. The heat absorbed by the cooling medium may be transferred for use in the process or for external supply. Water cooling is most often applied when waste LHV is high (e.g., > 12–15 MJ/kg of MSW).

Moreover, the use of water for grate cooling may allow grate metal temperature and local combustion temperature to be controlled with greater independence from the primary air supply. A better control of grate temperature can allow combustion of higher calorific value waste without the normally increased operational and maintenance problems. A typical waste feeder and grate device is shown in Fig. 3.3.

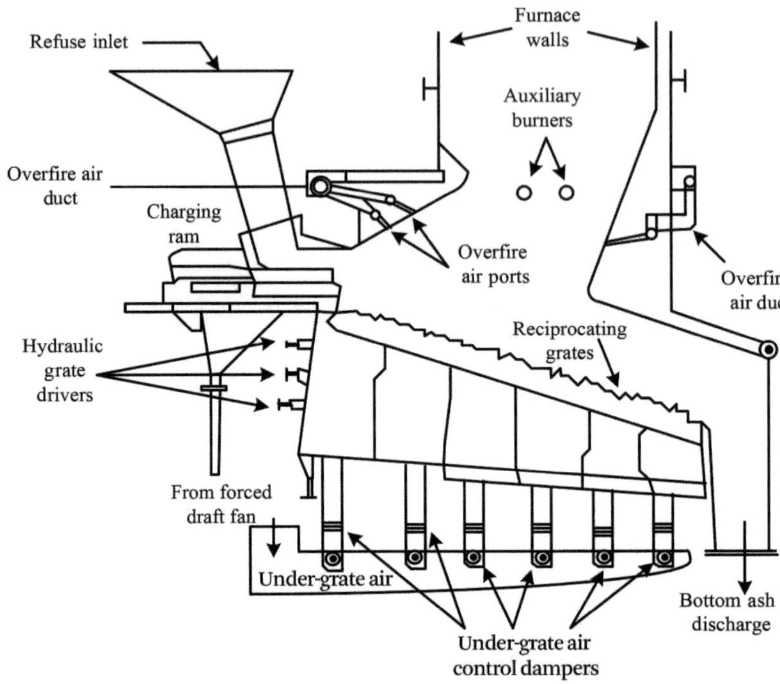

Fig. 3.3 Schematic of waste feeder, grate device, and combustion air system

Waste residence time over the grates is typically not more than 60 min. Different grate systems can be distinguished by the way the waste is conveyed through the different zones in the combustion chamber. There are essentially two main reciprocating grate variations, namely, continuous or discontinuous feeders:

- *Reverse reciprocating grate:* The grate bars oscillate back and forth in the reverse direction of the flow of the waste, consisting of fixed and moving grate steps.
- *Push-forward grate* (Fig. 3.4): The grate bars consist of many steps oscillating horizontally, pushing waste in the direction of the ash discharge section.

Combustion takes place above the grate in the incineration chamber (Fig. 3.5). As a whole, the incineration chamber typically consists of a grate situated at the bottom, cooled and uncooled walls on the furnace sides, and a ceiling or boiler surface heater at the top [1]. As municipal waste generally has a highly volatile content, the volatile gases are driven off, and only a small part of the actual incineration takes place on or near the grate. The following requirements influence the design of the incineration chamber:

- Form and size of the incineration grate: The size of the grate determines the size of the cross-section of the incineration chamber.
- Vortex and homogeneity of flue gas flow: Complete mixing of the flue gases is essential for good flue gas incineration.

Fig. 3.4 Forward-moving reciprocating grate

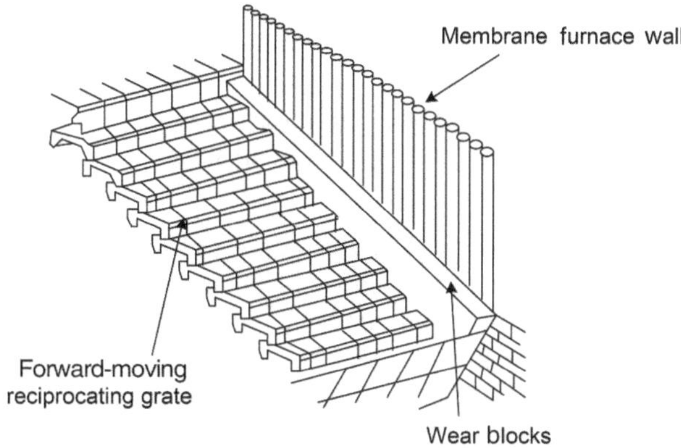

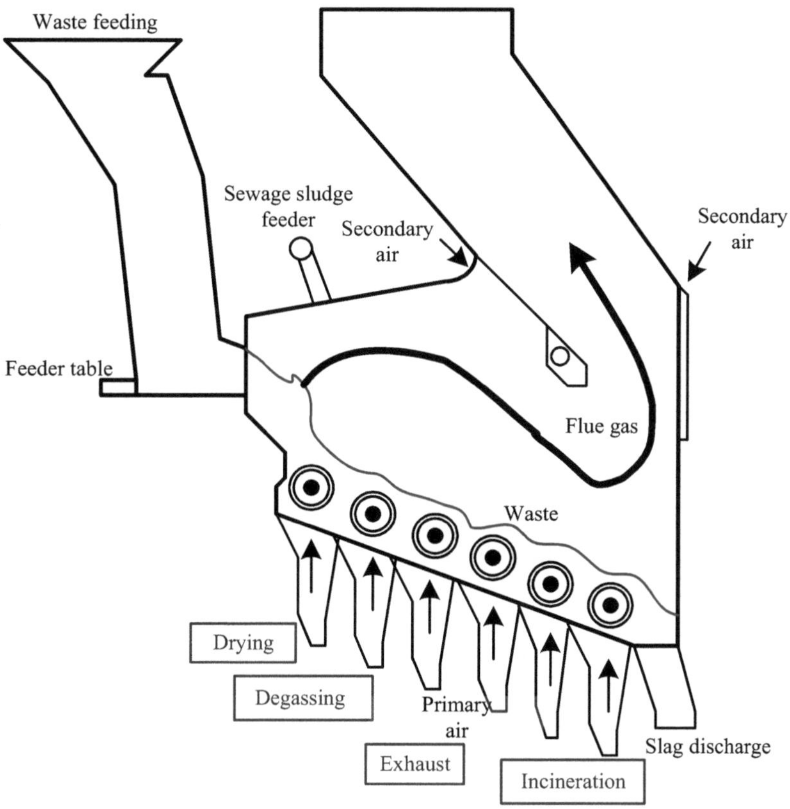

Fig. 3.5 Schematic of an incineration chamber

- Sufficient residence time for the flue gases in the hot furnace.
- Sufficient reaction time at high temperatures must be assured for complete incineration.
- Partial cooling of the flue gases in order to avoid fusion of hot fly ash at the boiler; the flue gas temperature must not exceed an upper limit at the incineration chamber exit.

Modern standards require flue gases to be exposed to a temperature of minimum 850 °C for a minimum residence time of 2 s after the last secondary air injection [1]. Therefore, the furnace must have a post-combustion chamber above the grate. The final burnout of the flue gas takes place in that chamber, and the required amount of secondary combustion air is added in a way that ensures maximum turbulence. To ensure complete combustion, a certain amount of excess combustion air is needed. The excess air, representing a loss of energy and hence a loss of thermal efficiency, can be reduced by using a water-cooled grate and/or by FGR. A fraction (approximately 10–20 % by volume [1]) of the (usually cleaned) flue gases is recirculated, normally after pre-dusting, to replace a fraction of secondary air feeds in the combustion chamber. This technique is reported to reduce heat losses with the flue gas and to increase the process energy efficiency by around 0.75–2 %. Additional benefits of primary NO_X reduction are also reported [1].

3.1.3 The Energy Recovery Section

Boilers for WTE power plants are typically water tube boilers, and most often they have four passes: three vertical radiation passes and a convective pass. The first radiation pass is integrated in the furnace (integrated boiler) as the post-combustion chamber. The convection pass, in which the evaporators, superheaters, and economizers are located, may be vertical or horizontal.

Thus, as schematically shown in Fig. 3.6, it is possible to differentiate three different heat surface areas:

- *Economizer* (or feedwater-preheating heat exchanger (HE); ECO in Fig. 3.6): In this area, the boiler feedwater is heated by flue gases to a temperature close to the boiling point.
- *Evaporator* (EVA in Fig. 3.6): In this area, the water coming out from the economizer is heated until it reaches the saturated steam temperature (designed as a bundled heating surface, envelopment wall of the incineration chamber).
- *Superheater* (SH in Fig. 3.6): In this area, the saturated steam coming out from the evaporator is superheated to the maximum temperature. Spray coolers and surface coolers are used in circulation boilers in order to maintain the exact required steam temperature. Their function is to balance the fluctuations of the steam temperature, which are the consequence of load fluctuations, changes in waste quality, surplus air, as well as contamination of the heat surfaces (clean or dirty surface).

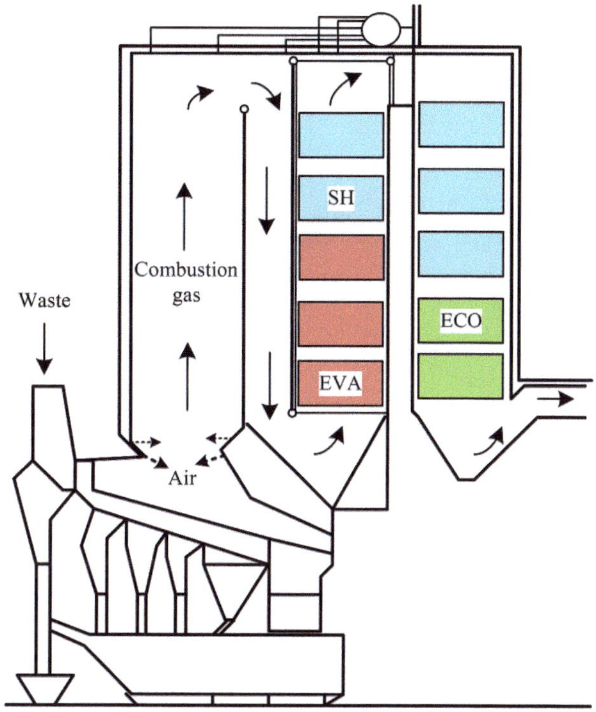

Fig. 3.6 Typical heat surface areas in a steam generator. *SH* superheating, *EVA* evaporation, *ECO* economizing

When designing a boiler for waste firing, the most important aspect to take into account is the special risk of corrosion. Corrosion problems, as described in detail in the following section, limit the steam parameters to a maximum of around 450–500 °C and 40–60 bar, respectively, for steam-superheated temperature and pressure. In general, at the entrance of the convection section (located after the third pass), flue gas has a temperature lower than 700–650 °C.

Good quality of the boiler and the adoption of a specific protective layer must be expected when increased steam parameters are used. Thus, a compromise between high steam cycle parameters and increased corrosion problems is required in order not to penalize waste conversion efficiency into energy and, at the same time, not to increase capital and maintenance plant costs.

Different boiler concepts can be used in waste incineration plants [1]:

- Horizontal boilers
- Vertical boilers
- Combination of vertical and horizontal boilers

Horizontal arrangement boilers are characterized by the fact that the flue gas in the convective heating surfaces travels horizontally. By contrast, in vertical arrangement boilers, the flue gas in the convective heating surfaces travels in the vertical direction. The main advantages of the horizontal arrangement are: (i) easier cleaning of the heating surface, (ii) reduced risk of tube bundle blocking due to particles

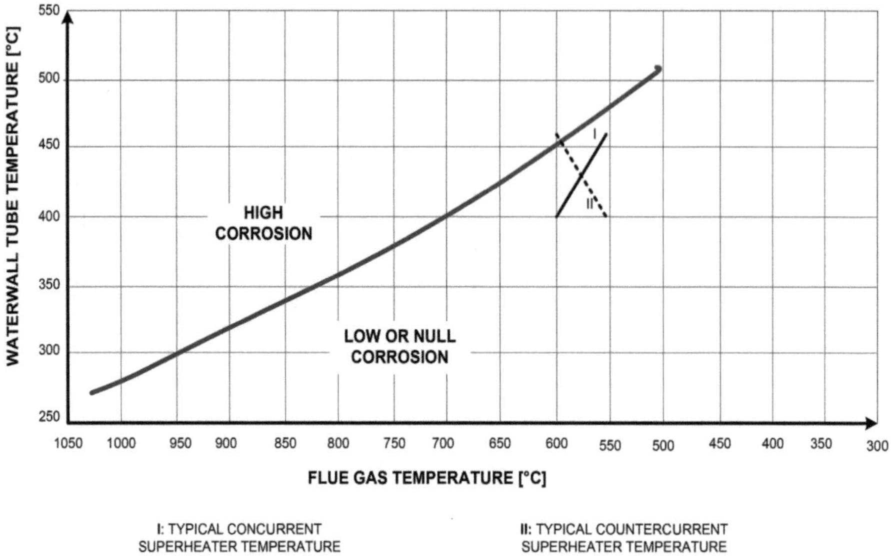

Fig. 3.7 Typical corrosion diagram

and dusts,[1] (iii) the possibility to place outside flue gas passing supports for the heating surface. But the major drawback of the horizontal configuration is that it needs huge space. Compactness is, however, the main advantage of vertical configurations, thus optimizing performance per metric tons of steel in the boiler.

3.1.3.1 Corrosion Protection

Modern WTE plants have significantly improved and are far superior compared with the older polluting incinerators. However, the variability in MSW LHV and its relatively high content of chlorine contribute to a highly corrosive atmosphere that shortens the life of the heat exchanger tubes [3]. The incineration chamber, the waterwalls of the first passes, and the superheaters are boiler components most affected by corrosion. A typical corrosion diagram, derived from practical experience, is shown in Fig. 3.7 [4]. The diagram, where horizontal and vertical axis report flue gas and waterwall surface temperature, respectively, highlights two different zones characterized by different corrosion rates. As highlighted in the figure, flue gas and waterwall tube temperature are the fundamental variables to determine corrosion risk.

[1] The high proportion of ash in flue gas causes a risk of a correspondingly high contamination of the heat transfer surfaces. This leads to a decline in heat transfer and therefore a performance loss. Thus, heat transfer surface cleaning plays an important role. This cleaning can be accomplished manually or automatically with lances (compressed air or water jet), with agitators, with soot blowers using steam, with a hail of pellets (sometimes shot cleaning), with sound and shock waves, or with tank cleaning devices.

The most important factors affecting corrosion probability inside a WTE boiler are [3]:

- *Metal surface temperature*. High temperature of the metal surface due to high radiation fluxes and/or inadequate heat transfer rate to the steam flow inside the tube results in the melting of deposits and acceleration of the corrosion rate. In general, the metal temperatures of waterwall and superheater tubes are maintained at temperatures below 300 and 450 °C, respectively. However, as mentioned earlier, operation at higher superheater temperatures increases the thermal efficiency of the steam turbine.
- *Gas temperature*. The temperature of the combustion gases can affect the deposition rates and also the composition of the deposit, and thus accelerate corrosion. The temperature gradient between gas temperature and metal surface temperature is a driving force for the condensation of vaporized species, such as metal chlorides, on the cooled surface.
- *Temperature fluctuation*. The heterogeneous MSW composition and the corresponding fluctuation in LHV with time results in significant fluctuations of the gas temperature within the combustion chamber. Experimental studies have confirmed that the corrosion rate increases several times because of wide temperature fluctuation.
- *Characteristics of molten salt deposits*. The presence of chlorides, sulphides, alkaline, and heavy metal components in deposits affects both the chemical and physical properties of deposits, such as gas permeability. Deposit composition is affected by waste feed composition and the gas–metal temperature gradient. Corrosion rate also increases with an increase in thickness of deposits.

A detailed description of corrosion mechanisms inside a WTE facility lies beyond the scope of the present work (additional information can be found in [3–9]).

Metal tube corrosion is the major operating problem as it results in downtime and periodic shutdowns in WTE plants and accounts for a significant fraction of the total operating costs of WTE plants. Moreover, as introduced in Sect. 3.1.3, the increase in steam cycle parameters to achieve high plant efficiency causes heat transfer surfaces to be exposed to a severe corrosion atmosphere.

Besides corrosion problems, another negative aspect of a WTE plant is erosion, that is, the abrasion of surface material through vertical wear-and-tear, which is primarily caused by the ash particles present in flue gas. Erosion appears mostly in the area of gas redirection. Tube wear is caused by a combination of corrosion and abrasion.

Countermeasures, only helping to reduce corrosion damage to an acceptable level, can be divided into primary and secondary measures.

Primary measures seek to eliminate corrosion by influencing the process conditions in the boiler [7], mainly minimizing gas temperature fluctuations and enhancing mixing of gas through gas recirculation and optimal boiler system design.

Secondary measures involve the use of protective layers to extend the lifetime of the boiler tubes. To prevent corrosive attacks, heating surfaces in the radiant section and superheaters are, typically, protected by resistant refractory materials and/or

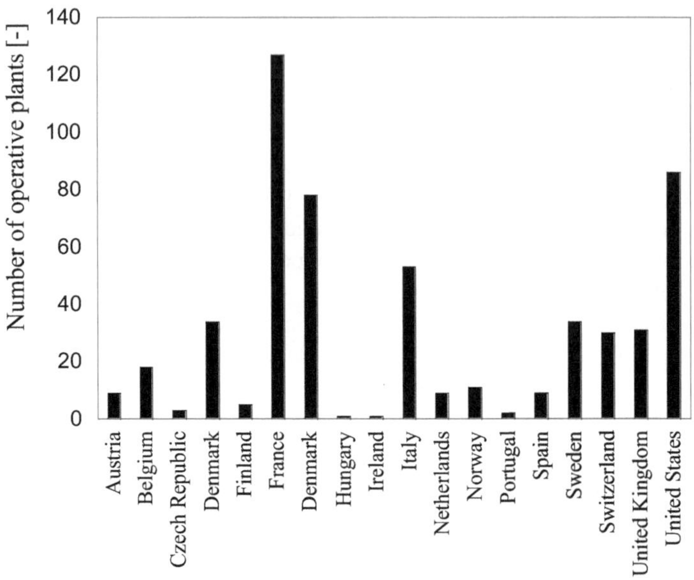

Fig. 3.8 Number of operative incinerator plants in EU-17 and the USA in 2011

welded high-alloy materials. Cladding consists of overlaying a layer of Inconel 625 on tubes which provides excellent corrosion resistance [8, 9]. The most important drawback in the use of Inconel 625 is its extremely high cost, which is only partially compensated by absent corrosion costs.[2]

A Waste-to-Energy Research and Technology Council (WTERT) study [10] shows that the nonscheduled downtime due to corrosion typically ranges from 0 to 20 days per year and that the yearly maintenance cost per boiler unit due to corrosion ranges from US$ 18,000 to 1,200,000 or from US$ 0.23 to 8.17 per ton of MSW combusted.

3.2 WTE Plant Distribution in the European Scenario

The International Solid Waste Association (ISWA) report [11] examined the situation in 18 countries (17 European countries and the USA) in 2011. Figure 3.8 shows the results of the study: The total number of operative WTE power plants in the European Union (EU)-17 was 455, while there were 86 operative WTE plants in the USA. The waste treatment capacity of each installation varies greatly across Europe: WTE plant capacity for each country is shown in Fig. 3.9. The average EU-17

[2] The application of Inconel 625 on superheater tubes is more complicated because the performance of the cladding depends on the metal temperature reached during operation [8, 9].

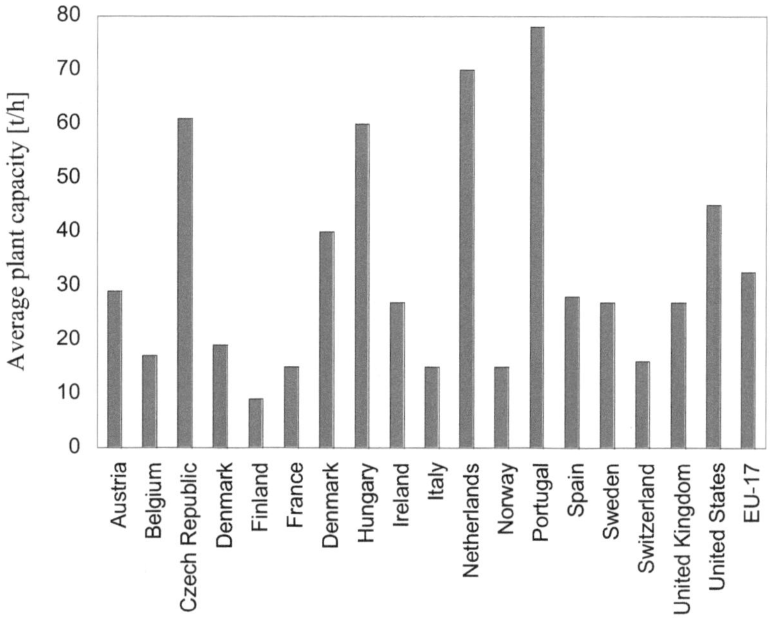

Fig. 3.9 Average incinerator plant capacity (t/h) in EU-17 and the USA in 2011

capacity value is 32 t/h; although in some countries very large plants are common, in others small plants are characteristic.

A recent study by the Confederation of European Waste-to-Energy Plants (CEWEP) [12] forecasts a significant increasing trend for incineration with energy recovery in Europe: The MSW treatment capacity of European plants will grow from 64 million metric tons in 2006 to over 100 million metric tons in 2020 (+59%), with a marked increase in the recovery of energy, both in the form of heat and electricity.

3.2.1 WTE Plant Efficiency in a Representative National Scenario

Focusing on Italy, as representative case, latest recorded data (2010) [13] show 50 operating incinerators with a total capacity equal to 7.12 Mt/year, corresponding to 2925 MW_t thermal capacity and to 782 MW electric power output. Energy recovery is obtained in almost all plants (49 out of 51[3]): All plants produce electricity, while only 11 plants are combined heat and power (CHP) systems. Figure 3.10 shows operative incinerators in Italy from 1995 to 2010, and Fig. 3.11 shows the distribution of WTE plant nominal gross electric size and thermal capacity. It is interesting to

[3] 51 incinerators are installed out of which 50 are operative plants.

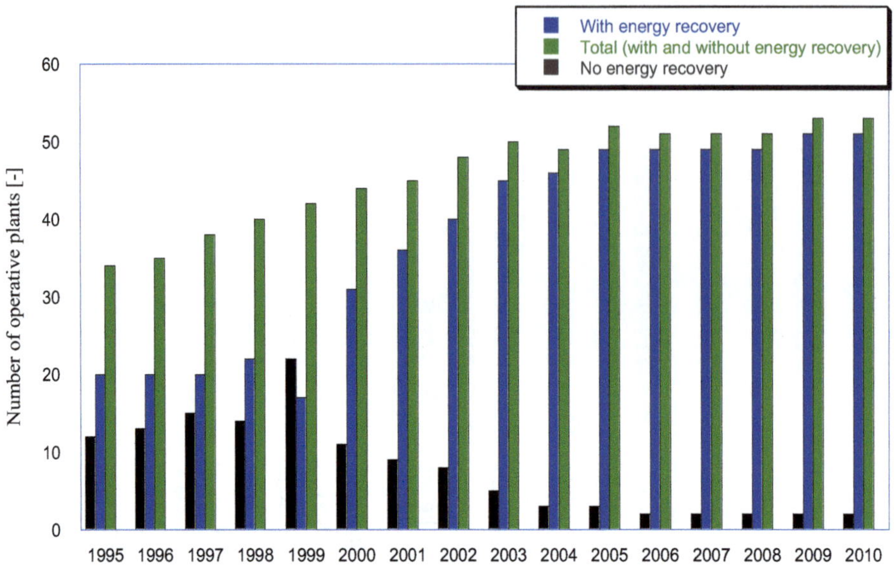

Fig. 3.10 Operative incinerators in Italy with and without energy recovery,1995–2010

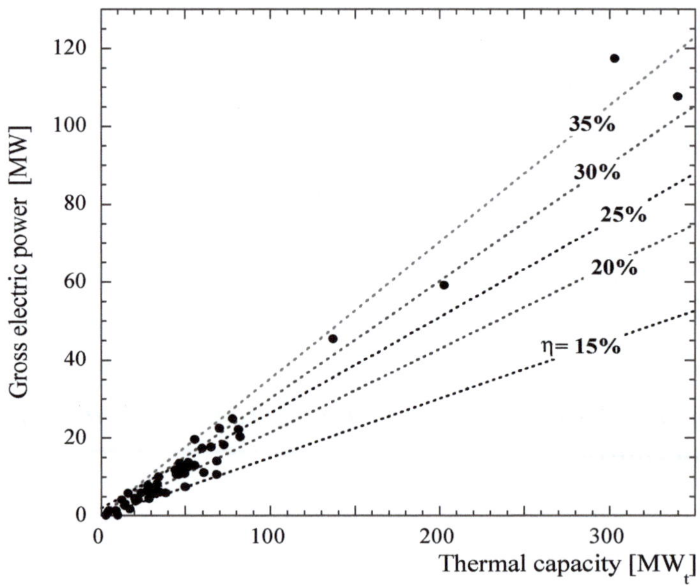

Fig. 3.11 WTE electric power output and input thermal capacity in the Italian scenario

note that 34 out of 50 plants have a capacity equal to or below 50 MW_t (mainly old generation), while only few new plants exceed 100 MW_t. Figure 3.11 also shows different values of gross electric efficiency, depending on the technology level, size, and age of the plant. In most cases, it is between 18 and 25%, while only in a few

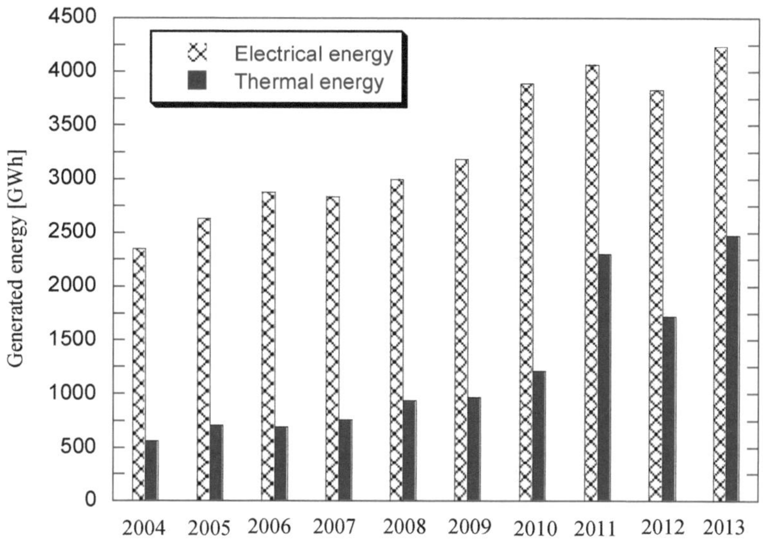

Fig. 3.12 Electrical and thermal energy generated in WTE power plants in Italy, 2004–2013

cases the gross efficiency is above 30 %. A detailed description of the state-of-the-art WTE steam cycle arrangement and parameters is carried out in Chap. 4.

Figure 3.12 shows the total amount of electrical and thermal energy generated by WTE power plants in Italy between 2004 and 2013 [13]. A growing trend of generated energy is evident, with an increase of more than 1888 GWh and 1913 GWh_t for electrical and thermal energy, respectively. A further significant increase in capacity is expected in the Italian scenario by the end of 2014.

3.3 EU Regulation Framework Oriented to WTE Efficiency

The 2008/98/EC WFD promotes production of energy from MSW, introducing an energy efficiency criterion (the so-called R1 formula [14, 15]) for WTE plants to obtain the "recovery" or "waste disposal" status under certain conditions. The aim of the WFD is to drive WTE facilities to maximize energy recovery and utilization of waste heat in order to substitute energy produced with conventional fossil fuel power plants.

The R1 ratio is a modified efficiency index, including in its expression all energy input/output of an WTE plant, that is, electricity and thermal production, losses, input waste, imported electricity, and additional fuels. The R1 formula [15] is defined as:

$$R1 = \frac{E_{\text{prod}} - (E_{\text{f}} + I_{\text{imp}})}{f_{\text{B}} \cdot (E_{\text{f}} + E_{\text{w}})}, \qquad (3.1)$$

where:

- E_{prod} is the annual energy produced as heat or electricity. It is calculated with energy in the form of electricity being multiplied by 2.6 and heat produced for commercial use multiplied by 1.1. So, E_{prod} can be regarded as an "equivalent primary energy": the energy input, with conventional fossil fuels, to generate the same amount of electricity and heat as the WTE power plants. It is important to understand that the equivalence values are not exact coefficients or conversion factors. They provide an estimation of the energy required to produce the same amount of energy output externally.
- E_f is the annual energy input to the system from fuels contributing to the production of steam (e.g., natural gas used, if necessary, as an additional fuel to increase post-combustion temperature).
- E_w is the annual energy contained in the waste, calculated using the net calorific value (LHV) of the waste.
- I_{imp} is the annual energy imported, excluding E_f and E_w.
- f_B is a factor accounting for energy losses due to bottom ash and radiation (assumed equal to 0.97).

A minimum value of R1 is required for WTE facilities to achieve the status (R1 = 0.60 for existing plants, R1 = 0.65 for new plants). It poses no problem for the few large, new and most advanced WTE plants to achieve the "energy recovery" status; by contrast, for the diffused, small and old facilities, only the less performing "waste disposal" status can be achieved. The reason is linked to the limited thermodynamic conversion efficiency of typical WTE power plants. As described in Sect. 3.1.3 [16, 17], low steam cycle parameters and small capacity of WTE facilities negatively affect the efficiency of MSW energy conversion.

Figures 3.13 and 3.14 show the maximum, minimum, and average R1 values for existing European WTE plants [12] and the thresholds for energy recovery status. In particular, Fig. 3.13 shows the effects of the plant capacity on the achieved R1 values; Fig. 3.14, however, shows a comparison of the achieved R1 values by WTE plants producing only electricity, or providing only heat, or by CHP plants. On average, R1 is higher for WTE plants with highest capacity (>250 Mt/year) which work in CHP mode.

The described scenario justifies a design trend towards large WTE plants with a CHP arrangement in order to achieve the energy recovery status. Moreover, even if R1 threshold values can be satisfied, this does not imply that high conversion efficiency is achieved in conventional WTE plants. Indeed, the current technological limits of conventional WTE plants, based on a superheated steam thermodynamic cycle, still penalize the electric energy efficiency of existing WTE plants.

Improved electric energy efficiency of MSW-fired steam units can be achieved only by tolerating higher corrosive rates for increased superheating temperatures, and consequently higher maintenance costs [16].

Thus, especially for small, inefficient but diffused WTE plants, new strategies to maximize the MSW energy conversion are currently under investigation. This calls for novel approaches and possibilities. In particular, the idea of an integrated

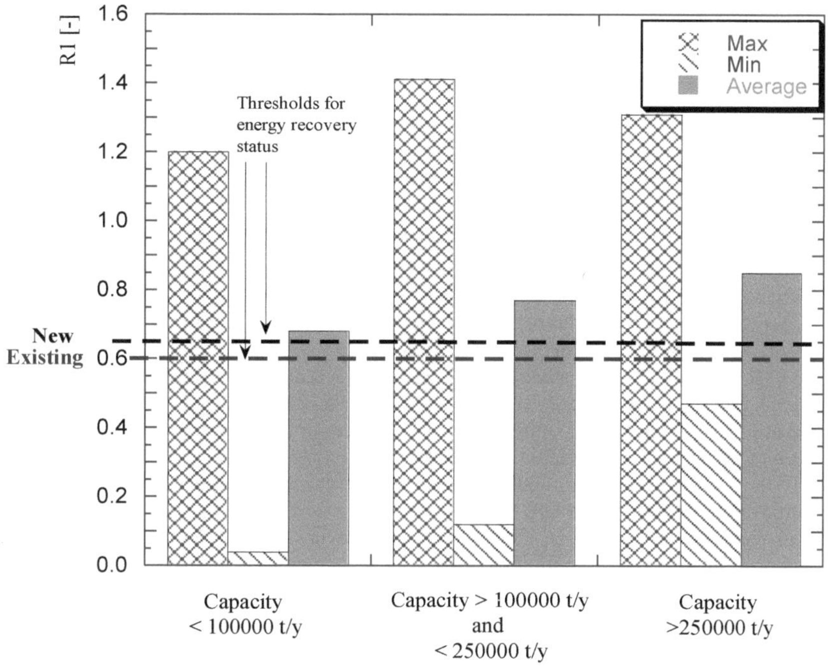

Fig. 3.13 Maximum, minimum, and average R1 values for existing WTE power plants as functions of plant size

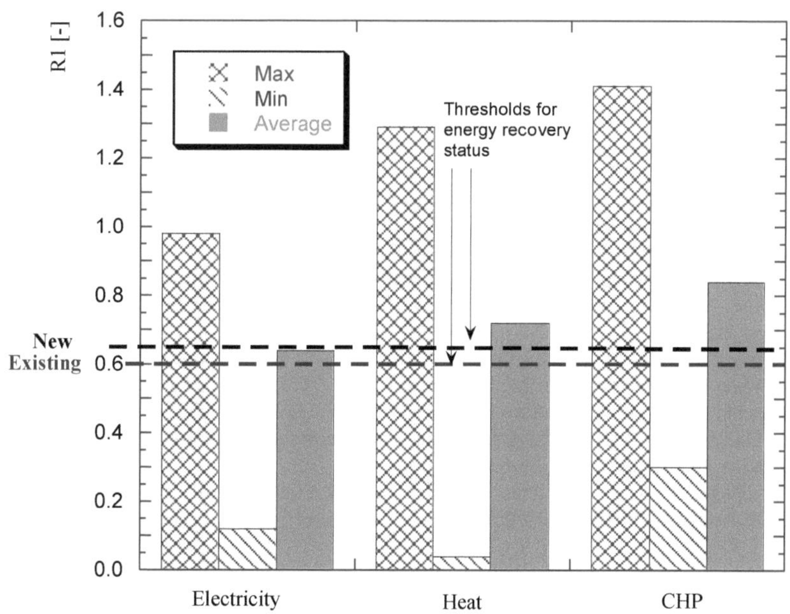

Fig. 3.14 Maximum, minimum, and average R1 values for existing WTE power plants generating only electricity, only heat, or working in the combined heat and power (*CHP*) mode

configuration made up of a WTE and a gas turbine (GT) originates from the desire to eliminate or, at least mitigate, limitations of the thermodynamic cycle affecting the WTE conversion process.

References

1. European Commission (2006) Integrated Pollution Prevention and Control Reference Document on the Best Available Techniques for Waste Incineration
2. Babcock & Wilcox volund brochure, 21' Century Advanced Concept for Waste-Fired Power Plants. http://www.volund.dk. Accessed March 2012
3. Lee SH, Themells NJ, Castaldi MJ (2006) High temperature corrosion phenomena in waste to energy boilers. J Therm Spray Technol 16(1):104–110. doi:10.1007/511666-006-9005-4
4. Carminati A (2003) Caldaie alimentate con RSU Accorgimenti per ridurre la corrosione, La Termotecnica – Novembre 2003. (In Italian)
5. CESI—Ricerca di Sistema (2005) Indagine problematiche di affidabilitÃ degli impianti di termovalorizzazione, A5049283
6. Rademarkers P, Hesseling W, Wetering J (2002) Review on corrosion in waste incinerators, and possible effect of bromine. TNO Industrial Technology
7. Albina DO (2005) Theory and experience on corrosion of waterwall and superheater tubes of waste-to-energy facilities. Master Thesis, Columbia University
8. Kawahara Y (2002) High temperature corrosion mechanisms and effect of alloying elements for materials used in waste incineration environment. Corros Sci 44(2):223
9. Wilson A, Forsberg U, Noble J (1997) Experience of composite tubes in municipal waste incinerators. Corrosion 97:153
10. Zwahr H (2003) Ways to improve the efficiency of waste to energy plants for the production of electricity, heat and reusable materials. Proceedings of the 11th North America Waste to Energy Conference, Florida
11. International Solid Waste Association ISWA (2012) Waste to energy state of the art report-statistics 6th Edition. http://www.iswa.org/fileadmin/galleries/Publications/ISWA_Reports/ISWA_Report_2012.pdf. Accessed March 2014
12. Reimann DO, CEWEP Energy Report II (status 2004–2007) Results of specific data for energy R1 plant efficiency factor and Net calorific value of 231 European Waste to Energy plants
13. ISPRA Istituto Superiore per la Protezione e la Ricerca Ambientale Edizione (2014) Rapporto Rifiuti Urbani, ISBN: 978-88-448-0665-1. http://www.isprambiente.gov.it/files/pubblicazioni/rapporti/RapportoRifiutiUrbani2014_web.pdf. Accessed 19 Sept 2014
14. Bianchi M, Branchini L (2012) Advanced waste-to-energy power plant: integration with gas turbine. International Conference on Applied Energy ICAE2012, July 5–8, 2012, Suzhou, China, Paper ID: ICAE2012-A10330
15. Directive 2008/98/EC of the European Parliament and of the Council of 19 November 2008 on waste. Official Journal of the European Union OJ L 312/3,22.11.2008
16. Bianchi M, Branchini L, Melino F, Peretto A (2010) Cycle upgrade for Waste-to-energy power plants, ASME-ATI-UIT 2010 Conference on Thermal and Environmental Issues in Energy Systems 16–19 May, 2010, Sorrento, Italy
17. Bianchi M, Branchini L, De Pascale A, Falchetti M, Fiore P (2014) Advanced waste-to-energy steam cycles. Energy Procedia 45:1205–1214. doi:10.1016/j.egypro.2014.01.126

Part II
WTE Thermodynamic Analysis

Chapter 4
Waste-to-Energy Steam Cycle

This chapter focuses on state-of-the-art waste-to-energy (WTE) steam cycle arrangement and possibilities to improve thermodynamic efficiency, upgrading both steam cycle parameters and cycle layout.

Based on a collection of WTE facilities data, a "state-of-the-art" WTE power plant is identified: Typical values of steam cycle pressures and temperatures are discussed with reference to the corrosion problem and characteristic plant size. Starting from the "state-of-the-art" WTE power plant, a sensitivity analysis is carried out to analyze the effects of the most important steam cycle parameters and configuration, improving the thermodynamic efficiency of the WTE facility. An initial assessment of the importance of the main steam cycle parameters must be outlined as a preliminary study in order to understand the possibilities and the benefits of a thermodynamic cycle upgrade for a WTE power plant.

4.1 Steam Cycle State-of-the-Art Parameters and Layout

The amount of energy recovered from waste combustion can vary significantly with the characteristics of municipal solid waste (MSW) fed into the boiler (composition; mass flow rate; and lower heating value, LHV), the combustion technology, the configuration and features of the recovery boiler (adiabatic or integrated), and the characteristics of the thermodynamic cycle.

For boiler normal operative conditions (excluding start-up and shutdown), as explained in Chap. 3, the kinetics of the corrosion process is influenced by the wall heat exchanger and flue gas temperatures. Thus, the characteristics of the produced steam, pressure, and temperature play a fundamental role in corrosion.

All of these factors expose the boiler structure to a variety of critical events; it is therefore necessary to protect the heat exchanger section with refractory materials, reducing the heat flux and limiting the flue gas temperatures, and to plan a specific heat exchanger section organization.

L. Branchini, *Waste-to-Energy*, DOI 10.1007/978-3-319-13608-0_4

The abovementioned problems concerning the WTE conversion limit the maximum superheated steam temperature. Aggressive compound condensation is another limiting factor. In order to avoid it, high outlet temperatures must be carried out.

These limits reduce the thermal energy input with waste actually convertible into useful energy, constraining the steam cycle efficiency.

Corrosion problems bind the inlet turbine temperature to be between 370 and 450 °C. Low superheating temperatures also imply moderate evaporation pressure in order to limit the liquid fraction at the steam turbine (ST) outlet, typically between 40 and 50 bar. Besides, higher values of the evaporation pressure (more than 50 bar) make a protection of the combustion chamber with a noble metal layer to avoid corrosion problems necessary.

It has to be pointed out that the possibility to increase the evaporation pressure and steam-superheated temperature is mostly limited by economical and plant complexity aspects.

Typical flue gas outlet temperatures range from 180 to 250 °C, significantly higher than those typical for fossil fuel power plants. Values equal to or below 160 °C can only be achieved in advanced plants.

The typical size of an incineration plant is smaller than that of a conventional fossil fuel plant. The electric power generated from the combustion of MSW (usually between a few megawatts up to a maximum of 70–80 MW) is one or two orders less than that of a conventional power plant. Moreover, the layout of the plant must be as simple as possible to reduce the investment costs.

Surface condenser with air-cooling towers or air cooler condenser is generally used for steam condensation, implying high condensation pressure (between 0.2 and 0.1 bar), while 0.05 bar is the value for conventional power plants using surface condensers with open cycles.

The net electric efficiency is also negatively influenced by the high air excess necessary for the MSW combustion and, as a consequence, by the high volume of combustion products. Thus, considerable amount of exhaust flue gases are discharged, raising the auxiliary power consumptions and the discharged heat.

Due to these reasons, the conversion efficiency of the steam cycle for a WTE power plant hardly exceeds 30 %, while the net average value of electric efficiency is around 25 %.

Figure 4.1 shows the distribution of Italian WTE facilities in terms of electric power output as a function of waste thermal input and steam cycle parameters (steam maximum pressure and temperature).

As stated by this study, evaporative pressure values range between 10 and 75 bar; 55 % of the cases (27 plants) are within a range of 20 and 40 bar, with a peak around 40 bar. The operating temperature is typically 400 °C. Brescia WTE power plants achieved the best performance with 75 bar and 450 °C of evaporation pressure and steam maximum temperature, respectively [1–3].

The average values of plant capacity and steam parameters, highlighted in Fig. 4.1, suggest that a WTE representative "average" facility can be identified. This representative "average" WTE is characterized by medium size (average input capacity equal to 55 MW_t and gross electric power output of 15.7 MW), moderate

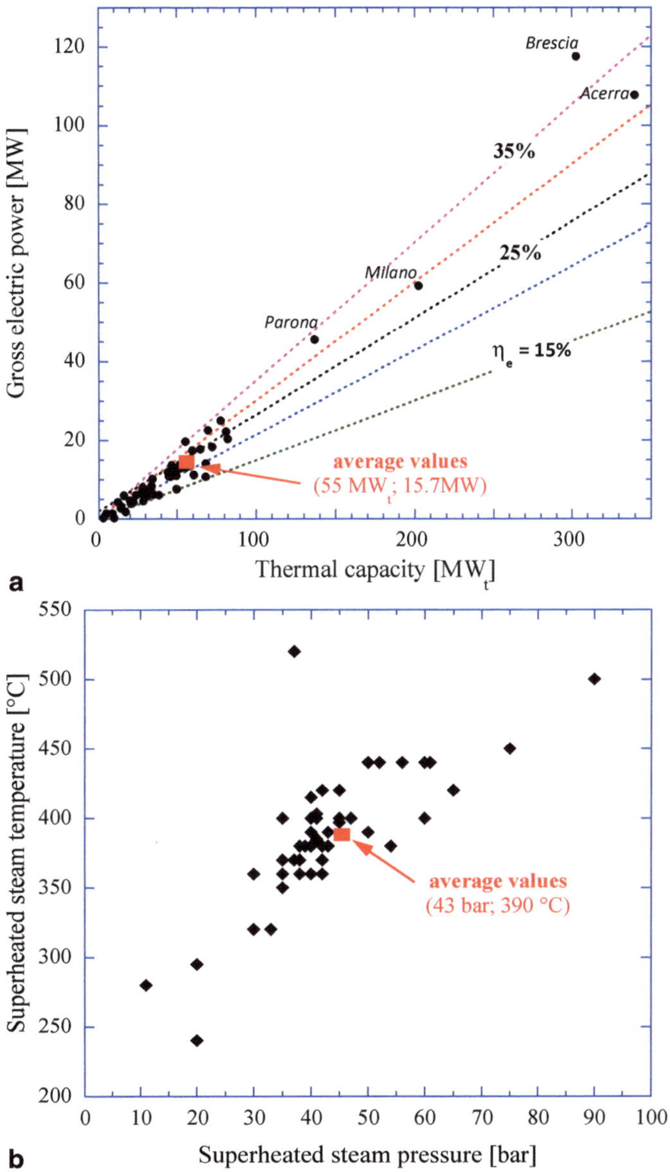

Fig. 4.1 Italian WTE gross electric power output and input thermal capacity (**a**) and steam cycle parameters (**b**)

steam cycle parameters (ST inlet pressure of 43 bar and temperature of 390 °C), and typically steam extraction from the turbine and regeneration.

Opposed to state-of-the-art "average" midsize WTE (Fig. 4.1), "high-performing" WTE power plants can be identified, characterized by high steam cycle parameters

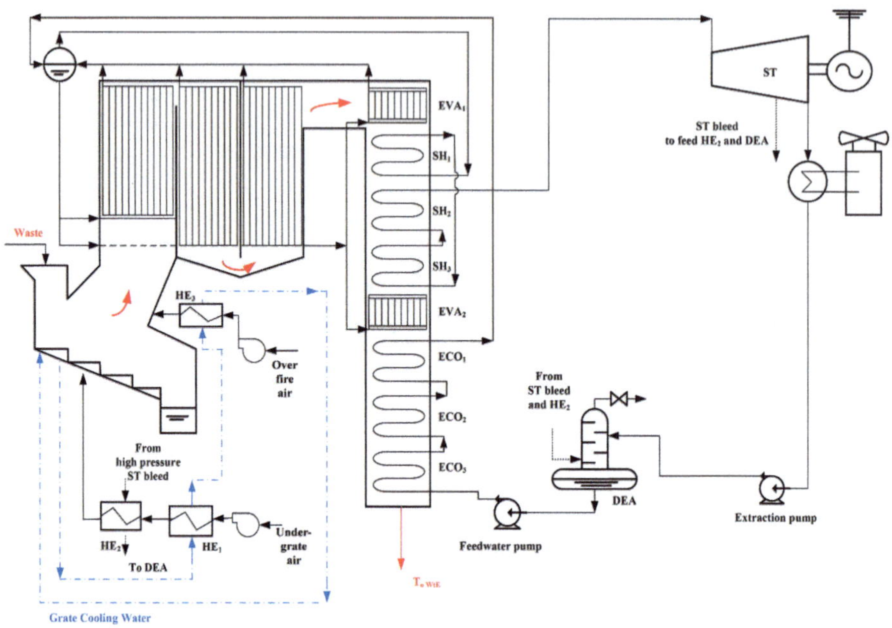

Fig. 4.2 Schematic layout of a midsize reference WTE power plant. *HE* heat exchanger, *ST* steam turbine, *DEA* deaerator, *EVA* evaporator, *SH* superheater, *ECO* economizer

(ST inlet pressure higher than 80 bar and temperature higher than 450 °C) and high plant capacity (over 100 MW_t). This high-performing steam cycle arrangement is typical of modern and high-size WTE power plants where significant investment costs have been accepted.

A superheated steam cycle without reheating and with steam extraction is the typical layout arrangement of a midsize WTE power plant. A schematic layout of state-of-the-art cycle arrangement is presented in Fig. 4.2. The steam generator is a natural circulation type, integrated with the combustion chamber. It comprises several radiation channels with vertical flue gas flow and a convection section containing the superheaters (SH), evaporators (EVA), and economizers (ECOs). As already detailed in Chap. 3, the configuration of the convective section can be either vertical, horizontal, or combining both types.

Flue gas produced by combustion of waste passes through the steam generator, where superheated steam required to be fed to the turbogenerator for the generation of electricity is produced. The SHs are tagged SH_1, SH_2, and SH_3, following the steam flow. The layout of SH banks, following the flue gas flow, is usually designed to limit corrosion from chlorine at high temperature; thus, high-temperature flue gas comes in contact first with SH_1, disposed in concurrent steam flow. The SH_1 parallel flow configuration is necessary to guarantee the lowest metal temperature because here the coolest steam conditions are exposed to the hottest gas temperature. Nevertheless, to obtain satisfactory refuse boiler SH performance, the use of Inconel tube material in the highest tube metal temperature sections is required.

While the second and third SH encountered by the flue gas are designed with countercurrent steam flow, carbon steel is commonly used in SH sections with lower metal temperatures [4].

The upper furnace, where EVA sections are placed, must be properly sized in order to provide adequate heat transfer surface to reduce the flue gas temperature entering the SH to an acceptable level (see also Sect. 3.1.3).

Superheated steam, generated inside the convective section of the WTE boiler in SH bank, is sent to the ST, connected to a synchronous generator for the production of electricity ECO. ECOs are located at the end of the WTE boiler convective section, while the EVA is split in different heat exchanger banks, located in both the radiating and the convective sections (EVA_1 and EVA_2) of the boiler.

ECOs can be either vertical longflow or horizontal crossflow.

An ST bleed is used to feed the deaerator (DEA) and to complete the primary air preheating before combustion. Air preheating is necessary for two reasons: (i) to help refuse-drying and ignition processes and (ii) to increase the thermal efficiency increasing the reactant's enthalpy. Primary (or under-grate) combustion air is typically preheated up to 120–150 °C. To achieve this temperature level, in addition to grate cooling exploitation an ST bleed feeding a heat exchanger is necessary. Secondary (or over-fire) air is generally preheated taking advantage just of grate cooling up to about 50°C.

Under-grate air systems are typically designed for 70 % of the total air with an expected normal operation of 60 % [4]. Because waste contains a high percentage of volatiles, a significant portion of the total combustion air should enter the furnace secondarily, or as over-fire air (about 40 % in normal operation), through the furnace wall. The basic function of secondary air is to complete combustion, providing both the necessary oxygen and the adequate turbulence necessary to mix combustion gases.

Water cooling with a closed loop is used for the combustion grate; after passing through the grate, hot water is used to preheat the primary and secondary inlet air streams. Thus, removed heat through grate cooling is exploited for combustion air preheating.

A fraction of exhaust gas (typically 15 % by volume) is recirculated inside the WTE boiler in the post-combustion section in order to reduce nitrogen oxide (NOx) emissions.

Minimum oxygen content, $\%O_2$, in exhaust gases from WTE boilers of 6 % was required by earlier legislation but was removed from the most recent European Commission (EC) directives on incineration [5]. Despite that, the typical operative value of oxygen content in dry exhaust gases ranges between 6 and 9 %. Lower O_2 content may increase carbon monoxide emission and may increase corrosion risk [5].

Minimum exhaust gas outlet temperature is mostly determined in order to avoid acid dew point corrosion. Minimum possible dry gas temperature is 130–140 °C using normal boiler construction material. When temperature falls below the acid dew point, wet chemical corrosions appear on cold surfaces. This damage can be avoided by raising the temperature or by selecting an appropriate material. Typical gas outlet temperature ranges between 150 and 190 °C for modern WTE facilities [5].

A second ST bleed (at lower pressure), if present, can be used in the feedwater heat exchanger (after extraction pump) to preheat water coming out from the condenser before entering the DEA. The presence of an additional ST bleed for feedwater preheating can be found in higher-capacity WTE, where additional complexity and associated costs can be more easily accepted.

4.2 Steam Cycle Upgrade: Effects on Cycle Efficiency

An initial assessment of the importance of the main steam cycle parameters must be outlined as a preliminary study in order to understand the possibilities and the benefits of a thermodynamic cycle upgrade for a WTE power plant. The selection of steam cycle parameters has a considerable impact on energy efficiency providing a great opportunity for an increased use of energy out of waste. Despite that, achievable conversion efficiency gain must be always weighed with increased investment cost that can occur.

A sensitivity analysis has been carried out to analyze the effect of the most important steam cycle parameters and configuration to improve the performance of a WTE plant. The parameters considered in this study are:

- Condenser pressure, p_k
- Evaporative pressure, p_{ev}
- Steam-superheated temperature, T_{SH}
- ST isentropic efficiency, η_{it}
- Exhaust gases temperature, $T_{O,\ WTE}$
- Oxygen content in dry exhaust gases, $\%O_2$
- Primary and secondary air temperature, T_{air}
- Cycle configuration in terms of regeneration

The influence of each key parameter has been studied considering a fixed capacity of the WTE power plant: MSW composition, mass flow rate, and LHV of waste fed into the integrated boiler are supposed to be constant.

If the input of thermal power with waste is constant, percentage increments in power output are achievable due to an upgrade of steam cycle parameters or the configuration and agree with increments in the electric efficiency value.

Thermodynamic and parametric analysis was carried out by the use of Gate Cycle™ [6], a commercial software from General Electric for plant design and simulation. The software, solving mass and energy balance using a lumped model approach, allows to evaluate inlet and outlet conditions of each system's components and to predict performance of the plant.

The influence of each key parameter has been studied individually or in relation to others.

Figure 4.3 shows the influence of evaporative pressure on electric efficiency for two different values of superheated steam temperature. Percentage increases achieved on efficiency are compared with a fixed increase in evaporative pressure

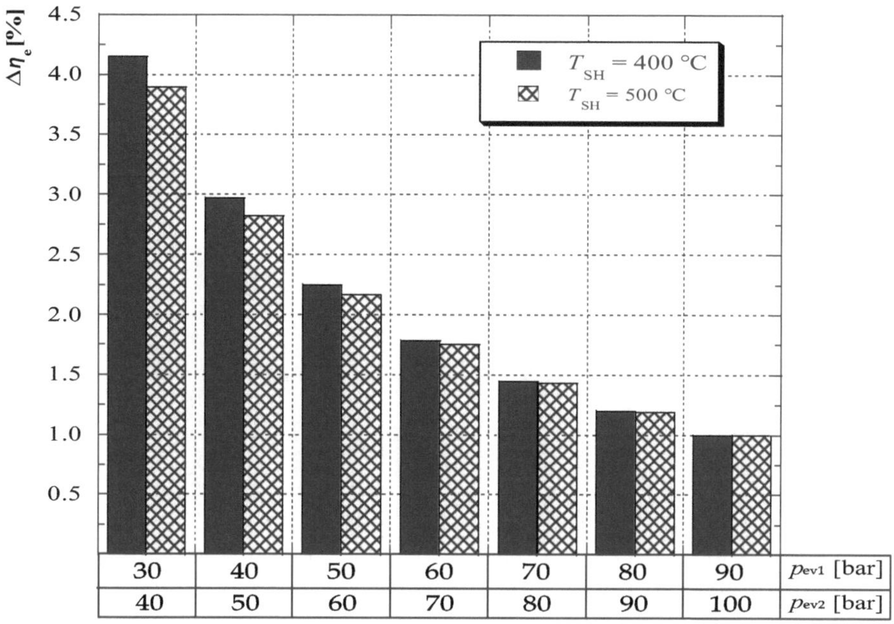

Fig. 4.3 Percentage increments on electric efficiency as a function of evaporative pressure for different values of steam-superheated temperature

value equal to 10 bar, that is, at 400 °C, passing from (p_{ev1}) 30 bar to (p_{ev2}) 40 bar; the increase in efficiency is equal to 4.2 %. This figure clearly highlights that higher benefits can be found for lower values of p_{ev}, passing from 4 % (from 30 to 40 bar) to about 1 % (from 90 to 100 bar). High values of T_{SH} slightly narrow the realized increase (Fig. 4.3).

Figure 4.4 shows the effect of condenser pressure on electric efficiency for different values of evaporative pressure. Percentage increments on efficiency are shown for a fixed decrease (equal to 25 mbar) in condenser pressure value. The possibility to reduce the p_k has considerable effects on the steam cycle efficiency. It should be noted that lower p_k values (i.e., 0.05 bar) can be achieved only by the use of surface condenser with an open cycle; this possibility is strictly limited by the presence of a natural water source close to the WTE power plant. Figure 4.4 clearly highlights that higher benefits can be found for lower values of p_k passing from more than 3 % (from 50 to 75 mbar) to about 1 % (from 1.75 to 2 mbar). Percentage gains increase if the evaporative pressure is lower (Fig. 4.4).

The increase in superheated steam temperature (T_{SH}) has an obviously positive effect on the efficiency of the steam cycle. The electric efficiency trend is shown in Fig. 4.5 for a fixed steam-superheated temperature increment, equal to 10 °C. The slope of the curve is slightly influenced by the actual value of T_{SH} (as evaporative pressure is equal to 100 bar); on the contrary, for a lower p_{ev} value, the percentage increment on efficiency increases with the increase of the actual value of T_{SH}.

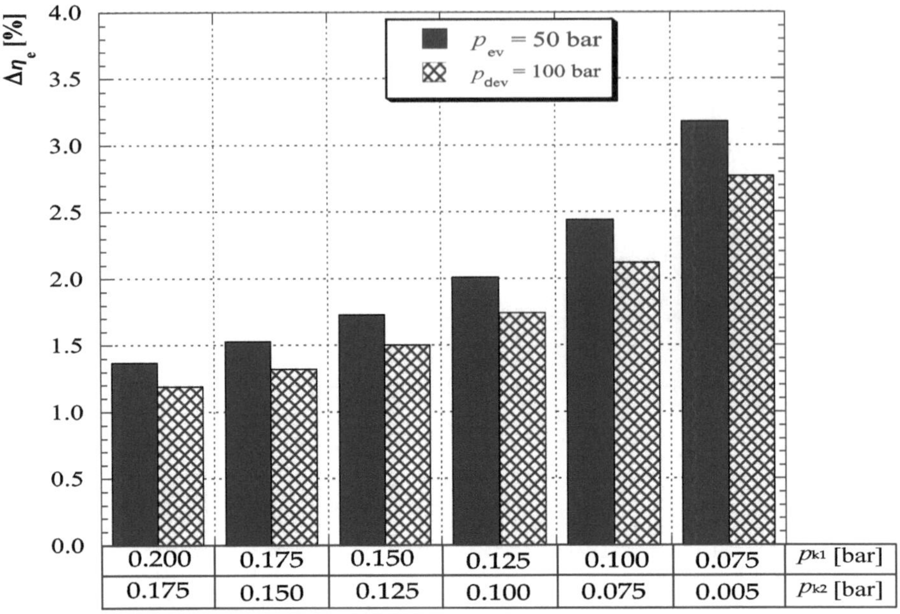

Fig. 4.4 Percentage increments on electric efficiency as a function of condenser pressure for different values of evaporative pressure

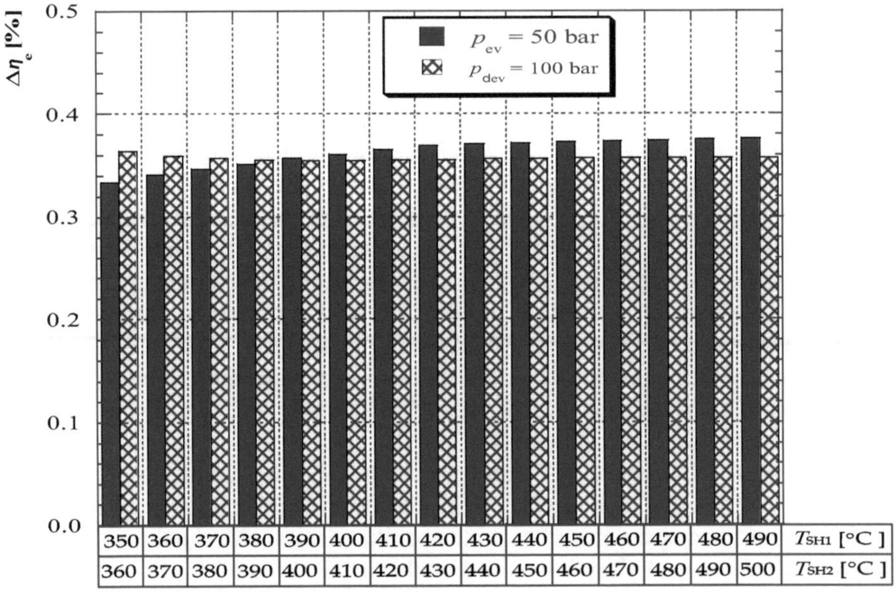

Fig. 4.5 Percentage increments on electric efficiency as a function of steam-superheated temperature for different values of evaporative pressure

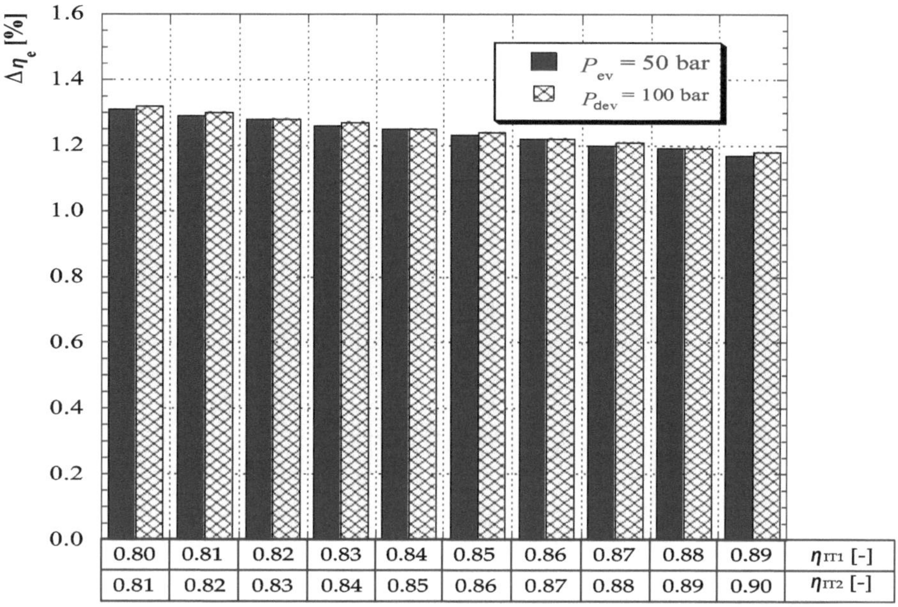

Fig. 4.6 Percentage increments on electric efficiency as a function of ST isentropic efficiency for different values of evaporative pressure

High values of ST isentropic efficiency (η_{it}) are also related to the size of the ST and, as a consequence, to the size of the plant. Percentage increments on electric efficiency are shown in Fig. 4.6 for a fixed η_{it} increment set equal to 0.01. Figure 4.6 points out that increments are not influenced by the evaporative pressure value.

In Fig. 4.7, percentage increments gained for a fixed decrease in $T_{O,\,WTE}$ equal to 10 °C are shown. Increments are almost constant, not dependent on evaporative pressure and steam-superheated temperature. It should be noted that, in the considered range, the lowest value (160 °C) is typical of advanced WTE power plants.

Even a reduction of oxygen content (vol%O_2) in the dry exhaust gases at WTE boiler outlet causes a reduction in the amount of heat lost to the stack. Benefits on efficiency are shown in Fig. 4.8 for two different values of evaporative pressure. Figure 4.8 emphasizes that the percentage gain for a fixed percentage point of decrease is a function of the actual value of %O_2 but is not dependent on the p_{ev} value.

The influence of secondary air temperature on electric efficiency has been also analyzed. A percentage increase in electric efficiency of more than 1.6 % has been found by preheating secondary combustion air from ambient temperature to 150 °C.

By combining both secondary air and feedwater preheating, a gain of about 5.5 % can be achieved on WTE electric efficiency.

The sensitivity analysis results for each analyzed steam cycle parameters are summarized in Table 4.1. Of course, cycle improvements can be also combined together to obtain the corresponding electric efficiency increase.

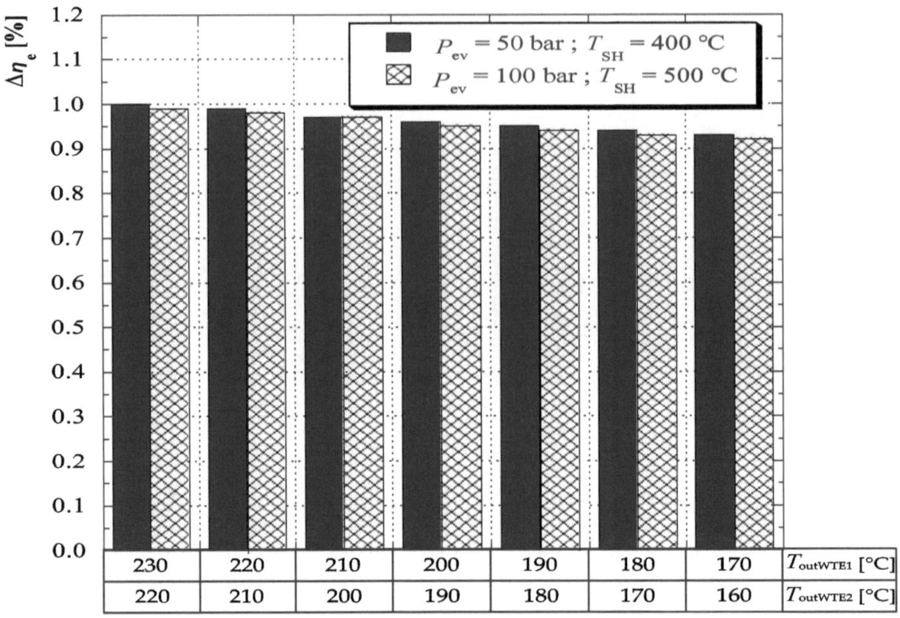

Fig. 4.7 Percentage increments on electric efficiency increments as a function of exhaust gas outlet temperature for different values of evaporative pressure

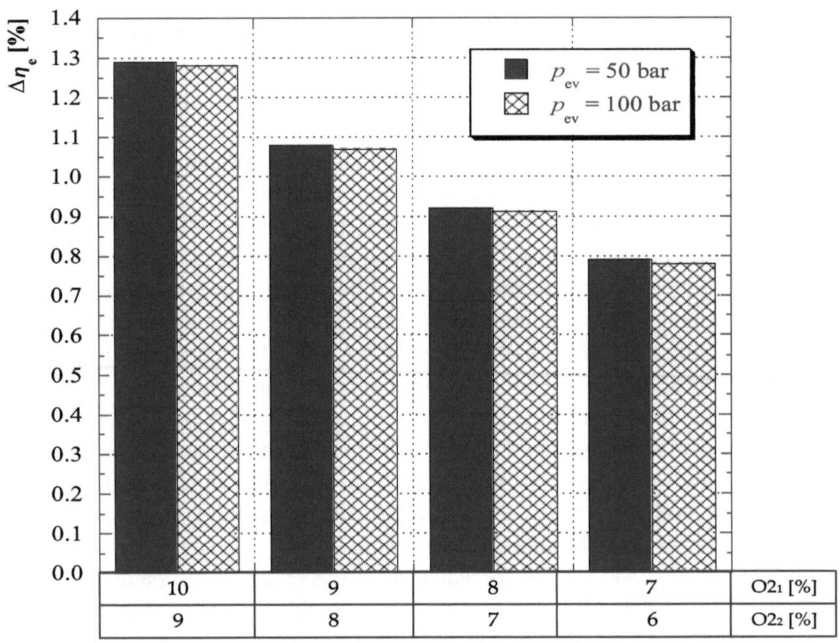

Fig. 4.8 Percentage increments on electric efficiency increments as a function of volume oxygen content in exhaust dry gases for different values of evaporative pressure

Table 4.1 Steam cycle improvements: effect on WTE efficiency

Steam cycle operative parameter	Assumed variation (increase or decrease)	Percentage increment gained on WTE electric efficiency (%)
Evaporative pressure, p_{ev}	+10 bar	1–4
Steam-superheated temperature, T_{SH}	+10 °C	0.35
Condenser pressure, p_k	−0.025 bar	1–3
ST isoentropic efficiency, η_{it}	+0.01	1.25
Exhaust gases outlet temperature, $T_{O,WTE}$	−10 °C	0.95
Oxygen content in dry exhaust gases, $\%O_2$	−1 %	0.8–1.3

WTE waste-to-energy, *ST* steam turbine

It has to be pointed out that some strategies, such as increasing evaporative pressure and steam-superheated temperature or decreasing O_2 content and exhaust gas temperature, require specific solutions to protect the integrity of the waste-fired boiler. Increased corrosion risks will follow both plant complexity and investment cost raises.

Along with upgrading efficiency of key steam cycle parameters, increase can also be achieved with cycle improvements. In fact, the steam cycle layout has a significant impact on WTE plant's performance. To better analyze the influence of a regenerative steam cycle on electric efficiency, the influence of regeneration for feedwater preheating has been investigated; results are shown in Fig. 4.9. An increase equal to about 5 % on efficiency can be obtained with preheating the water before entering the DEA. With an additional ST bleed, preheating the water before the ECO inlet will further raise the efficiency to over 22 %. As shown in Fig. 4.9, increments are quite independent of evaporative pressure values.

4.3 New Designs for High-Efficiency WTE Plant

Concerning high-efficiency WTE plants, some interesting cases must be discussed. New WTE concepts are still under investigation aiming at increasing plant efficiency. Proposed and already constructed plant featuring new design concepts are located in:

- Amsterdam—the Netherlands: *high-efficiency WTE plant*
- Reno—Nord Denmark: *SteamBoost* concept
- Germany—Karlsruhe Institute of Technology research: *bypass of fuel gas* concept

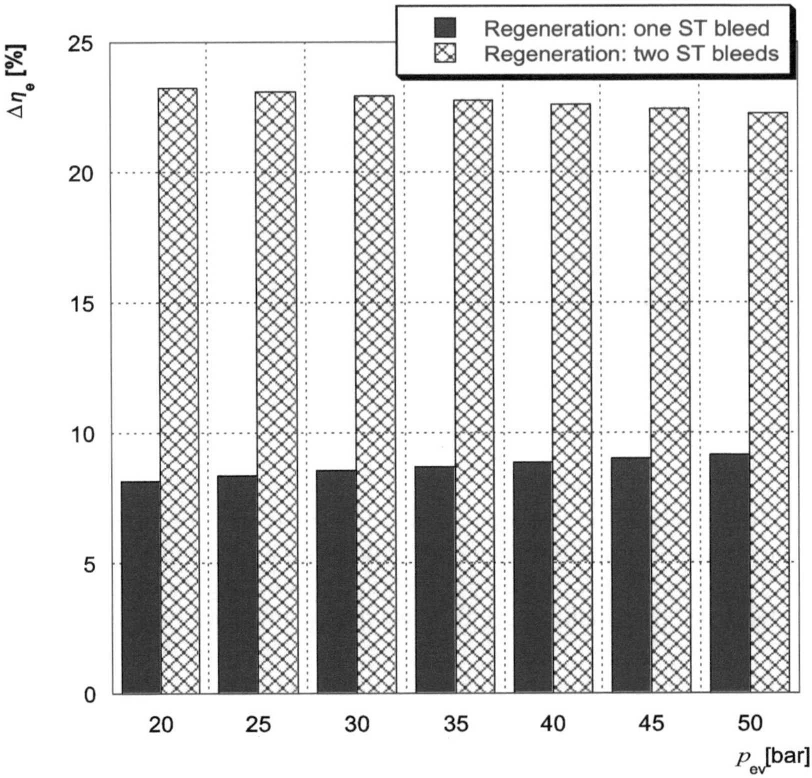

Fig. 4.9 Percentage increments on electric efficiency for regenerative cycles for different evaporative pressure values

The first highly efficient WTE plant, installed in Amsterdam, uses higher steam temperature/pressure together with an intermediate superheating using steam from the boiler drum and achieves a net electrical efficiency of 30 % [7]. The concept applied in Amsterdam involves, basically, the reheating of a part of the steam out of the high-pressure turbine using a steam-steam shell-and-tube heat exchanger to obtain a net electrical efficiency of 30 % and availability of over 90 % by using Inconel 625 as boiler cladding. Thus, a more classical approach is applied to WTE in order to increase efficiency as widely diffused in classical fossil-fueled steam cycles.

In Fig. 4.10, a schematic of the WTE concept with steam intermediate reheating is presented. As illustrated in Fig. 4.10, an intermediate reheater (RH), as additional component, is present where the saturated steam from the boiler drum is used to reheat the steam that comes out from the high-pressure ST.

In this plant, the steam pressure is increased from the typical 40 bar to 130 bar.

The temperature of the live steam at the outlet of the SH is 440 °C, while the temperature of the reheated steam is 320 °C. The main advantage of this configuration is its high energy efficiency due to the high pressure and the reheating combined with modest SH temperature of 440 °C.

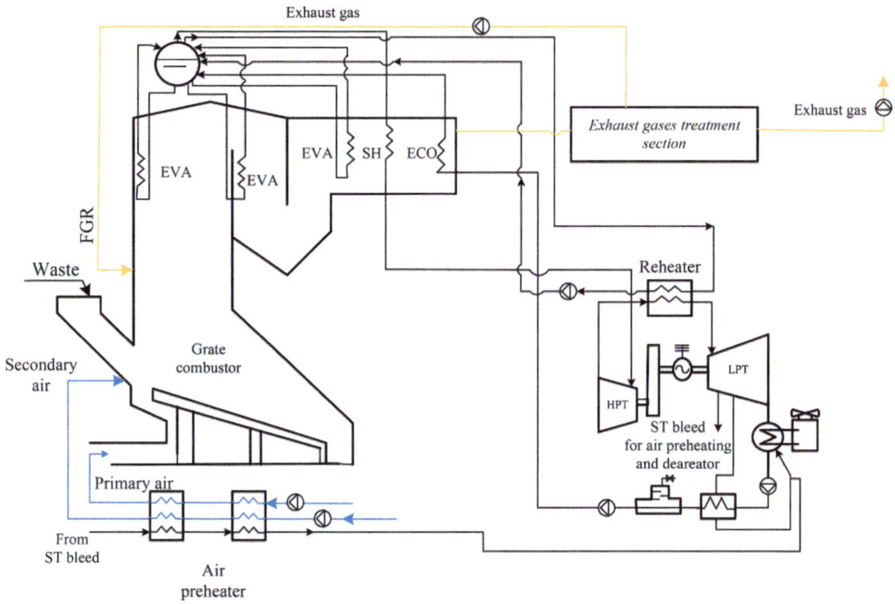

Fig. 4.10 Schematic of the Amsterdam WTE concept of steam intermediate reheating. *FGR* flue gas recirculation, *EVA* evaporator, *SH* superheater, *ECO* economizer, *ST* steam turbine, *HPT* high-pressure turbine, *LPT* low-pressure turbine

Combustion takes place at an excess air ratio equal to 1.4 in order to reduce flue gas losses. A portion of the flue gas is recirculated back into the lower part of the boiler, which reduces the temperature and improves the mixing in the post-combustion zone. In this plant, a condensing pressure of 0.03 bar is used (considerably below the typical WTE plant condensing pressure value).

Different concepts to further enhance the WTE plant efficiency of plants focus on flue gas and the fuel gas splitting concept, proposed by Babcock & Wilcox and Karlsruhe Institute of Technology Research Center, respectively.

More in detail, the Babcock & Wilcox Company has developed a new approach to improve the net electrical efficiency of WTE plants receiving a world patent [8]. The basic idea of the proposed concept (see Fig. 4.11) is to divide the flue gas from the grate into two fractions, one fraction of the flue gas having a highly corrosive content of chlorine and another fraction having a low chlorine concentration since the corrosibility of the flue gas varies significantly over the grate length. The low corrosive part of the flue gas may be directed to a separate SH section to increase the steam temperature, thereby boosting the electrical efficiency of the plant [8]. When using an extra SH with the flue gas split concept, the steam data are increased to 50 bar/500 °C, thereby increasing the efficiency by 3 % [9,10]. This method is applied in Denmark, in the Reno-Nord WTE CHP plant.

The Karlsruhe Institute of Technology Research Center has proposed a new technology for WTE plants based on the fuel gas splitting concept. In this configuration, schematically presented in Fig. 4.12, the temperature of the steam leaving the main

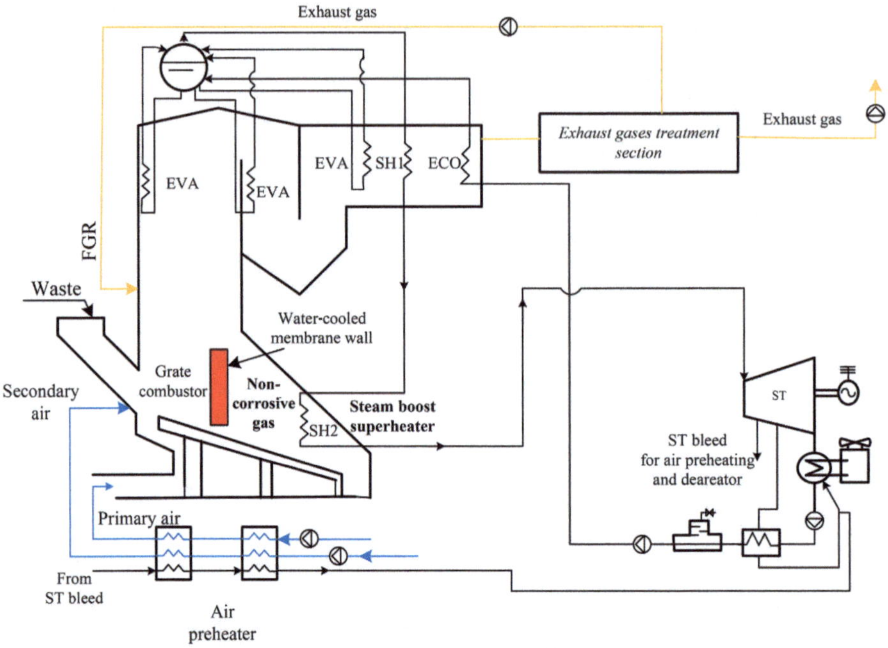

Fig. 4.11 Schematic of the Reno WTE concept of the steam boost concept. *FGR* flue gas recirculation, *EVA* evaporator, *SH* superheater, *ECO* economizer, *ST* steam turbine

boiler at 120 bar/400 °C is further increased up to 540 °C by burning the extracted fuel in a small combustion chamber and utilizing the generated heat for improved superheating of the pre-generated steam.

Combustion of MSW takes place in the main furnace, but a fraction of the formed fuel gas is extracted before entering the flue gas burnout zone and goes to a secondary combustion chamber, where high steam parameters are reached.

Therefore, bypass of a part of the fuel gas is achieved; this bypass system has the following components [7]: heat exchanger, fuel gas cleaning system, secondary combustion chamber, and SH.

As shown in Fig. 4.12, at the temperature of 900 °C, a part of the fuel gas is extracted from the furnace and subsequently cooled down to about 400 °C using a heat exchanger to preheat the air. In fact, the cooled gas has to be cleaned prior to combustion in the secondary combustion chamber in order to operate steam superheating at high temperatures avoiding corrosion risks.

At this temperature, alkali and heavy metal chlorides present in the fuel gas are transferred from the gaseous to the solid state; injections of alkaline compounds to the fuel gas allow efficient absorption of HCl and, to a lesser extent, of H_2S. Subsequently, the cleaned fuel gas is fed into the secondary combustion chamber of the bypass system. Here it is burnt with low air excess atmosphere. From low corrosive flue gas, generated heat is used in the superheater of the bypass system for enhanced temperature of the preheated steam streaming out of the furnace. It must be pointed

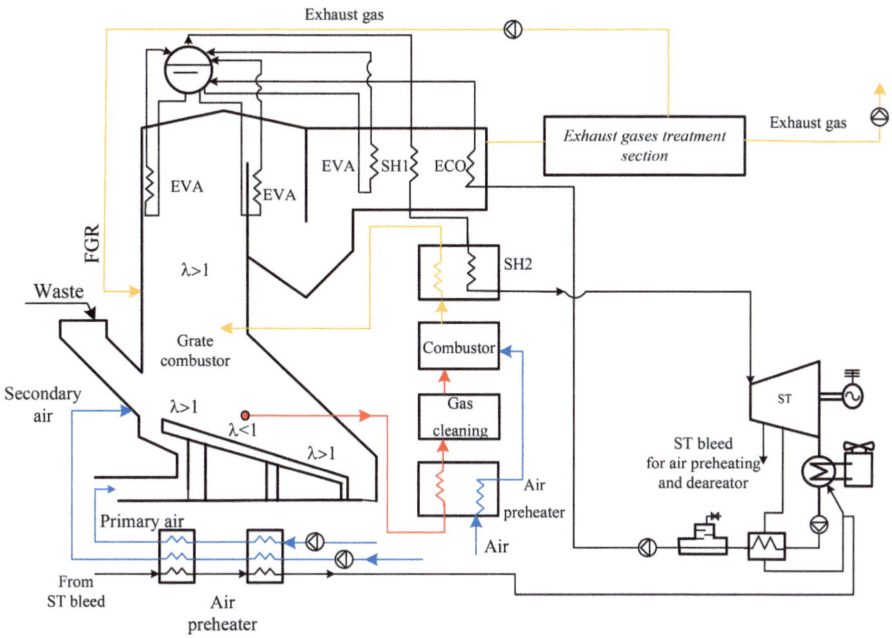

Fig. 4.12 Schematic of the Germany-Karlsruhe WTE concept of bypass of fuel gas [7]. *FGR* flue gas recirculation, *EVA* evaporator, *SH* superheater, *ECO* economizer, *ST* steam turbine

out that the external burnout of the fuel gas and the recirculation of the bypass flue gases into the main furnace allow the reduction of the excess air in the main combustion chamber by reducing the secondary air supply. As a benefit, the heat loss in the exhaust gas decreases, improving energy efficiency.

In this configuration, the temperature of the steam leaving the main boiler at 120 bar and 400 °C is further increased up to 540 °C. Compared with the typical WTE plant steam parameters, that is, 40 bar and 400 °C, the efficiency of gross power generation shows an increase of approximately 4 % points [7].

References

1. ENEA- FederAmbiente (2012) Rapporto sul recupero energetico da rifiuti urbani in Italia, 3° edizione, Marzo 2012 (in Italian).
2. Bianchi M, Branchini L, Melino F, Peretto A (2010) Cycle upgrade for waste-to-energy power plants, ASME-ATI-UIT, 2010, Conference on thermal and environmental issues in energy systems 16–19 May, 2010, Sorrento, Italy
3. Branchini L (2012) Advanced waste-to-energy cycle, PhD Thesis
4. Stultz SC, Kitto JB (ed) (1992) Steam its generation and use, 40th edition. The Babcock & Wilcox Company, Barberton, Ohio U.S. A., ISBN 0-9634570-0-4
5. European Commission (2006) Integrated pollution prevention and control reference document on the best available techniques for waste incineration, August 2006

6. Gate CycleTM by General Electric Energy (o J) http://site.ge-energy.com/prod_serv/products/oc/en/opt_diagsw/gatecycle1.htm
7. Bogale W, Viganò F (2013) A preliminary comparative performance evaluation of highly efficient Waste to Energy plants, ATI 2013–68th conference of the Italian thermal machines engineering association. Energy Procedia 45(2014):1315–1324. doi:10.1016/j.egypro.2014.01.137
8. Ralf K (2008) Innovative concepts of high-efficiency EfW plants, 16th annual North American waste-to-energy conference, May 19–21, Philadelphia, Pennsylvania, USA
9. Madsen O (2008) High electrical efficiency by dividing the combustion products", 16th annual North American waste-to-energy conference, May 19–20, Philadelphia, Pennsylvania, USA
10. Madsen O (2007) Next generation of waste fired power plants, NAWTEC 15, Miami, USA

Part III
WTE Advanced Cycles

Chapter 5
Waste-to-Energy and Gas Turbine: Hybrid Combined Cycle Concept

This chapter focuses on the hybrid combined cycle (HCC) concept. The HCC, based on thermal integration between a topper cycle (TC) and a bottomer cycle (BC), denotes specifically "dual-fuel" combined power cycles. The possibility to use different fuels for the TC and BC is one of the advantages of the HCC. This chapter describes in detail two basic types of hybrid dual-fuel combined cycle (CC) arrangements applied to waste-to-energy (WTE) power plants using as TC a gas turbine (GT), a steam/waterside-integrated HCC, and windbox repowering.

Furthermore, a literature review on HCC applied to WTE and GT is discussed, and three operative WTE power plants based on HCC distributed worldwide are described together with their technical data.

5.1 The HCC Concept

To introduce the concept of a waste-to-energy–gas turbine (WTE–GT) integrated power plant, a general overview on HCC is presented in this section.

Although standardized terminology to address the thermal connection between different systems does not exist, CC is the term commonly used for it. The CC concept is not new and widely spread: It dates back to as long back as the idea for a conventional gas-steam CC, based on GT as topper unit and a steam cycle as bottomer.

The term topper cycle (TC) addresses the power cycle of any heat engine that accepts thermal energy at a high temperature level and whose remaining exhaust heat is exploited in another cycle at a lower temperature level. Typical examples, widely spread today as TC, are GTs and internal combustion engines (ICE). TC systems typically utilize high-grade gaseous or liquid fuels.

The term bottomer cycle (BC) refers to any power cycle whose heat supply comes in the form of rejected heat from another power cycle. The BC itself rejects heat at the lowest possible temperature level. A typical example of a BC is the well-known Rankine cycle, working with water/steam or any other two-phase fluid

L. Branchini, *Waste-to-Energy*, DOI 10.1007/978-3-319-13608-0_5

(such as organic Rankine cycle). Power units of this type can have their own fuel input while serving as BC to a topping engine.

The term hybrid cycle denotes specifically "dual-fuel" combined power cycles where in different fuels are used for the TC and BC [1]. One of the main advantages of the HCC is the possibility to utilize low-grade fuels (solid fuels, such as municipal solid waste (MSW) or biomass) in the BC, together with exploiting the full potential of high-grade fuels (gaseous or liquid) in the TC.

The BC has then its own individual combustion chamber in which the bottoming fuel is fired. Two basic types of hybrid dual-fuel CC arrangements are possible (see Fig. 5.1) depending on the TC purpose:

- *Windbox repowering,* where the TC exhaust, with or without precooling, is supplied to the bottoming boiler and used as combustion air for firing the BC fuel
- *Steam/waterside integration,* where thermal energy from the TC exhaust is utilized for feedwater preheating and/or steam superheating and/or additional steam generation parallel to the BC

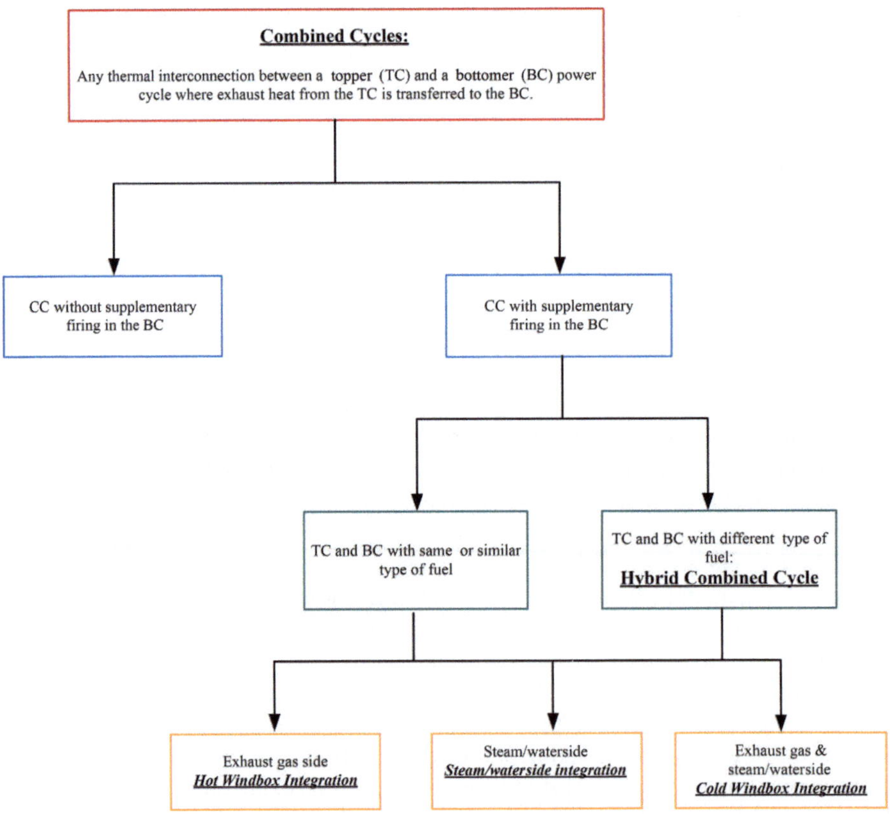

Fig. 5.1 Schematic classification of combined cycles and TC–BC integration options. *CC* combined cycle

Windbox repowering cycles can be further divided in two main categories: hot windbox and cold windbox. In the hot windbox type, the GT exhaust is fed directly into the BC boiler. In the cold windbox type, the GT exhaust is first cooled down to a lower temperature level (by various options, for example, supplying heat for parallel steam generation or feedwater preheating), after which it is fed into the BC boiler. Cold windbox arrangements actually allow for features typical to both the windbox repowering and steam/waterside-integrated hybrid cycles.

As will be discussed in detail, steam/waterside integration represents a cost-effective method to increase the electric efficiency in comparison with a reference WTE plant without new, expensive materials; indeed, further superheating of the steam is performed in a separate heat recovery steam generator (HRSG) by means of a high-grade fuel, which cannot cause corrosion problems in the superheater (SH) bank. By contrast, both cold and hot windbox repowering arrangements, due to a considerable increase in terms of combustion gas volumetric flow rate, require a significant redesign of the original WTE combustion, convective, and gas cleaning sections.

5.1.1 WTE–GT Steam/Waterside Integration

Focusing on steam/waterside integration, the WTE plant basically acts as an additional source of saturated steam and, if convenient, of hot water for the CC. All the equipment for power production is concentrated in the "island" comprising the CC. The waste treatment island comprises the equipment for waste handling, waste combustion, and flue gas treatment. Figure 5.2 schematically shows a possible WTE–GT steam/waterside-integrated power plant layout.

Being almost exempt from corrosion, the HRSG can take steam to temperatures typically reached in fossil fuel power plants (520–560 °C). In turn, higher steam temperatures allow higher evaporation pressures without undue liquid fractions in the last stages of the steam turbine (ST). Also, generation of saturated steam outside the HRSG (in the WTE boiler) gives a more favorable temperature profile (smaller ΔT between gas and steam or water) with lower irreversibility and higher efficiency.

The flue gas treatment section and thus pollutant concentrations at the stack can be identical to those of conventional WTE plants; this also applies to emissions specific to the amount of MSW because the integration causes no dilution of the combustion products. No dilution also means no penalties on the cost for flue gas treatment.

Increasing steam temperature externally in the Rankine cycle will improve the efficiency of electricity generation for MSW-fired power plants. This can be achieved by MSW and natural gas HCC that involve two different thermodynamic cycles with two types of fuel. In the CC, the TC consists of a GT, while the BC, driven by low-quality MSW fuel, is a steam cycle. In a dual-fuel CC system, there must be a well-designed thermal link between the TC and the bottomer steam cycle. The integration has to provide thermodynamic and operating advantages for both the TC and the BC. Generally, steam superheating by turbine exhaust heat is viable. This arrangement can substantially increase the efficiency of MSW energy conversion while avoiding the described corrosion problems (see Chap. 3).

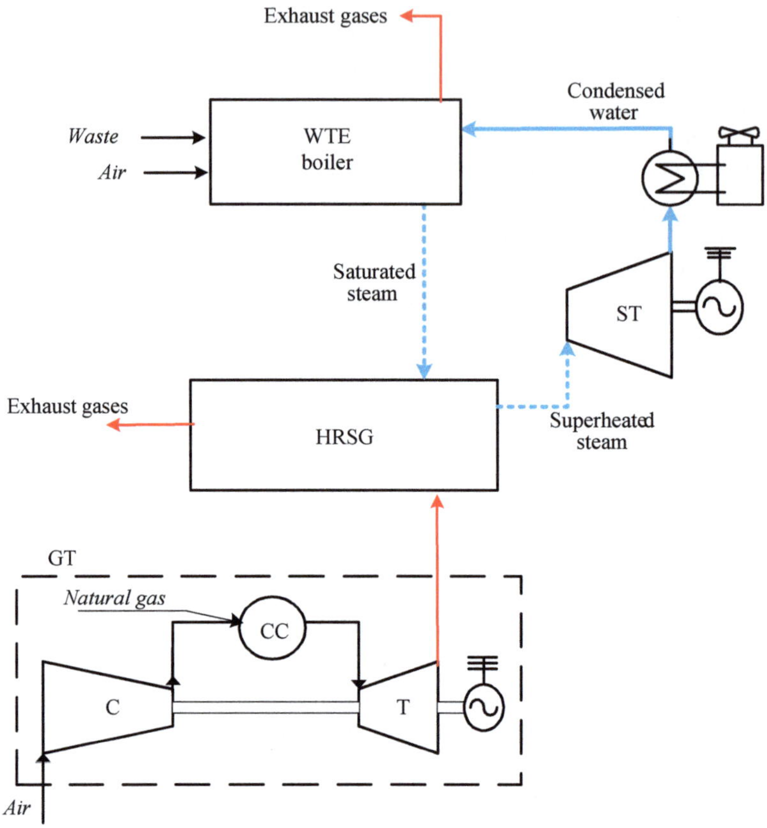

Fig. 5.2 Schematic of the steam/waterside integration between a waste-to-energy *(WTE)* and a gas turbine *(GT)*. *C* compressor, *ST* steam turbine, *HRSG* heat recovery steam generator, *CC* combined cycle, *T* turbine

The potential advantages of a WTE–GT plant match compared with a stand-alone WTE plant can be summarized as follows [2]:

- Increase in the maximum temperature of the steam-cycle-transferring SH from the WTE to the HRSG, the most problematic component for what concerns high-temperature corrosion.
- Reduction of the HRSG irreversibility caused by high mean temperature differences, moving the generation of saturated steam from the HRSG to the WTE section.
- Increase in ST efficiency due to its increased size: The ST serving both the WTE section and the HRSG is larger than that typically used in conventional WTE plants.

Moreover, from an economic point of view, the integrated system leads to further benefits such as the reduction of:

- Maintenance costs of the WTE section, due to elimination of the SH heat exchanger
- Capital costs since a significant number of equipment and services are shared in the integrated plant

5.1.2 WTE–GT Windbox Integration

Windbox repowering is accomplished by installing a GT to provide extra power and direct its exhaust into the original boiler's windbox with or without precooling.

Figure 5.3 shows a schematic WTE–GT hot windbox-integrated layout, while in Fig. 5.4 a schematic of cold windbox integration is presented.

GTs can be used for retrofitting existing steam plants (hot windbox repowering). GTs are operated with large amounts of excess air. Thus, the exhaust of GTs contains a high concentration of oxygen (14–16 vol%), and the exhaust temperature is also high (480–600 °C). This makes a GT well suited for integration with an MSW boiler in the CC [3]. Moreover, modern GTs are generally optimized with respect to maximum power density (output per unit air flow) rather than efficiency. This coincidentally meets well with the requirements of optimum efficiency of the CC plants.

The hot windbox repowering has the highest degree of technical complexity of all the combustion-turbine-based repowering options. The air heaters may need to

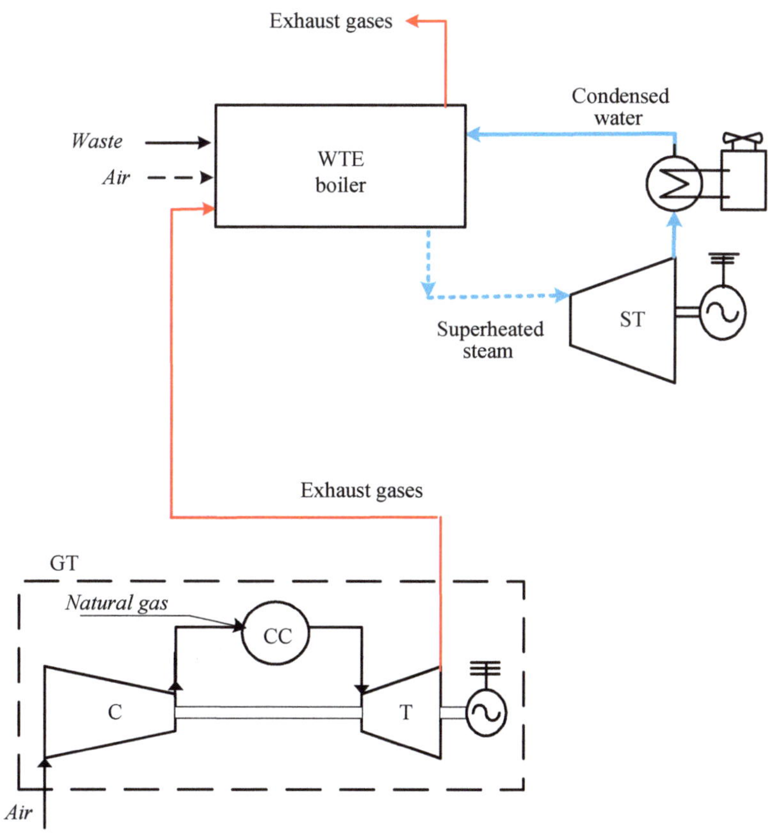

Fig. 5.3 Schematic of hot windbox integration between a waste-to-energy *(WTE)* and a gas turbine *(GT)*. *C* compressor, *ST* steam turbine, *CC* combined cycle, *T* turbine

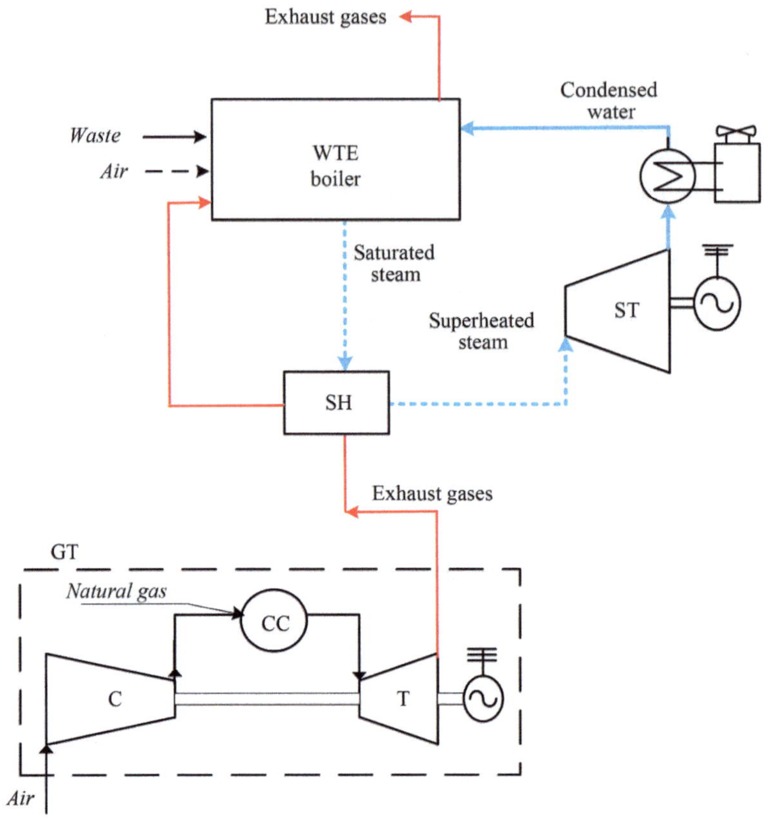

Fig. 5.4 Schematic of cold windbox integration between a waste-to-energy *(WTE)* and a gas turbine *(GT)*. *C* compressor, *ST* steam turbine, *ST* steam turbine, *CC* combined cycle, *T* turbine

be modified based on the revised air and gas flows, and the ductwork must be upgraded to accommodate the higher temperature and larger volume of air. The furnace burners must be modified or replaced because of the lower oxygen content of the flow from the combustion turbine exhaust. Furthermore, the lower oxygen content of the combustion air will change the heat release profile in the furnace, and some derating of the boiler or a redesign of the convective parts of the furnace may be necessary. Other necessary modifications can include bypass ducts for admitting variable amounts of combustion turbine exhaust, a steam air heater to allow independent operation of the existing boiler when the combustion turbine is not available, an induced draft fan to reduce the back pressure on the combustion turbine, and a combustion turbine bypass stack for unit start-up [4].

A variant of the hot windbox repowering approach includes a heat exchanger to reduce the temperature of the combustion turbine exhaust and, for example, produce additional steam. The possibility to share both gas and steam/water paths between cycles is referred to as "cold windbox repowering," presented in Fig. 5.4.

Advantages of hot and cold windbox WTE–GT integration can be summarized as follows:

- Elimination of WTE combustion air preheating due to the higher temperature of the GT exhaust gas.
- Reduction of the total exhaust gas mass flow rate and thermal power discharged in comparison with WTE and GT stand-alone systems.
- Reduction of the environmental impact compared with separate systems: GT exhaust gas would be subject to the same WTE exhaust cleaning treatments.
- Reduction of the water fraction in WTE exhaust gases that lower the acid dew point allowing a decrease in WTE minimum allowed temperature: The increase in mass flow rate in respect to fresh combustion air, for the same water content in the waste, decreases the water fraction in the WTE exhaust.

Performance of windbox integration can be further improved if external superheating is applied in the cold windbox configuration. In this case, the integrated cycle can have a twofold advantage: The steam SH located in the GT exhaust duct is not exposed to the corrosive gases and can thus achieve the same level as that in the SH, but it has a simpler design and a much smaller surface area than a heat recovery boiler.

5.2 State-of-the-Art of Integrated WTE–GT

HCC with MSW as BC fuel has become a very attractive topic for scientific and industrial research during the past 15 years [5–15] due to the growing interest in (i) energy utilization from waste and (ii) improvement in the electrical efficiency of WTE with cost-effective methods.

As already detailed in Chap. 4, improved electrical efficiency from MSW-fired steam units can be achieved by tolerating higher corrosive rates for increased superheat temperatures and consequently higher maintenance costs. A more cost-effective method for increasing the electrical efficiency without new materials or expensive investments is the possibility to superheat the steam in a separate heat exchanger by combusting cleaner fuels. The incorporation of the MSW boiler as a BC into an HCC, where the topping exhaust provides superheating, can substantially improve the MSW conversion efficiency into electrical energy using only conventional technology. This is relevant to any steam cycle with low steam parameters.

Several authors investigated possible arrangements of HCC plants with WTE. First, Eber et al. [5] proposed external superheating of steam, coming from an MSW boiler, in a simple natural gas burner. The study, where natural gas had been limited to 25 % of the total energy input, highlighted the thermodynamic and economic feasibility of this solution.

The idea of integrating WTE with a topping GT was put forward by Lowry and Martin [6] and Wiekmeijer [7]. Lowry and Martin evaluated a simplified arrangement in which the GT exhaust gas superheats the MSW-generated steam.

Wiekmeijer's study focused on a more complicated steam/waterside integration arrangement with an economizer (ECO) and a final SH in an HRSG [7].

Terasawa and Ogura [8] mentioned the HCC alternative for rationalization of MSW incineration practices. Ito et al. [9] evaluated the economic and energy characteristics of an MSW boiler integrated with a topping GT in combined heat and power (CHP) mode. Sue [10] suggested the application of a steam-injected GT as TC in an MSW-fired power unit. Otoma et al. [11] performed a life cycle analysis (LCA) for WTE with low steam parameters and an electric efficiency of 15 %. Then they evaluated options for topping this base-case boiler with a GT. Holmgren [12] focused on a thermodynamic and sensitivity analysis of the overall performance of three different HCC configurations of a WTE–GT in CHP mode. Results highlighted the advantages of the HCC in comparison with separate plants (existing MSW-fired WTE, CHP plant, and gas turbine combined cycle (GTCC) plant). Korobitsyn et al. [13] examined different WTE–GT HCC configurations, featuring final steam superheating by the GT exhaust gas. Their results showed an increase in the HCC electric efficiency compared with the average efficiency of the individual single-fuel plants (one simple MSW-fired steam cycle with efficiency close to 25 % and one GTCC). Moreover, this study suggested definitions to evaluate within the HCC the separated contribution of MSW to the electric efficiency. Finally, Udomsri et al. [14] investigated the economic performance of a WTE–GT, showing that HCC is a viable solution in comparison with conventional WTE.

Despite this large literature, the actual spread of integrated WTE–GT plants is really poor [15], confirming that this is a solution which is still new and under development. To the author's knowledge, only two WTE–GT plants are located in the European Union (EU; Zabalgarbi, Spain [16–18], and Moerdijk, the Netherlands [19]) and one in Japan (Takahama [20]). Only the first one has been designed as a new WTE–GT integrated plant while the others are retrofits of WTE. The limited spread of the integration concept can be mainly linked with the lack of regulations concerning multiple-fuel and renewable incentive allocation, rather than technical issues.

The most promising short-term HCC solution not requiring a complete redesign of the MSW boiler is the steam/waterside integration investigated in the following section. One important aim of this study not covered in the above cited references is to find out the optimal GT size for the proposed WTE–GT integrated configurations.

5.3 Existing WTE–GT Integrated Power Plants

The spread of integrated WTE–GT power plants is really poor, which confirms that this solution is new and still under research. There are two operative WTE–GT power plants in Europe, namely in Spain and in the Netherlands. Another operative plant is located in Japan. All the existing and operating WTE-GT power plants feature the steam/waterside-integrated configuration. However, it must be pointed out that only the Spanish WTE–GT integrated power plant was originally projected with the idea of an integrated power plant while the Dutch and the Japanese WTE power plants have been repowered by means of a GT as TC.

5.3.1 Zabalgarbi WTE–GT Power Plant: The SENER Solution

The integrated power plant of Zabalgarbi was built by Constructions Industrielles de la Méditerranée (CNIM), an expert in MSW with energy recovery construction and operation and SENER Ingeniería y Sistemas, S.A., an engineering company for the design and development of the project in Bilbao, Spain [16, 17]. Plant construction started in September 2001 and was completed after approximately 36 months. The start-up of the plant was in 2004.

The WTE–GT power plant is based on steam/waterside integration. It utilizes the SENER-2 high-efficiency thermodynamic energy cycle [17] designed and patented by SENER to minimize corrosion problems which are normal in conventional plants. As stated in the plant summary report, the aim of the project was to promote a new concept for generation of electricity from MSW to allow efficient power generation, efficient waste disposal, and a low environmental impact.

The process and a schematic layout of the plant are shown in Fig. 5.5, and technical data are summarized in Table 5.1.

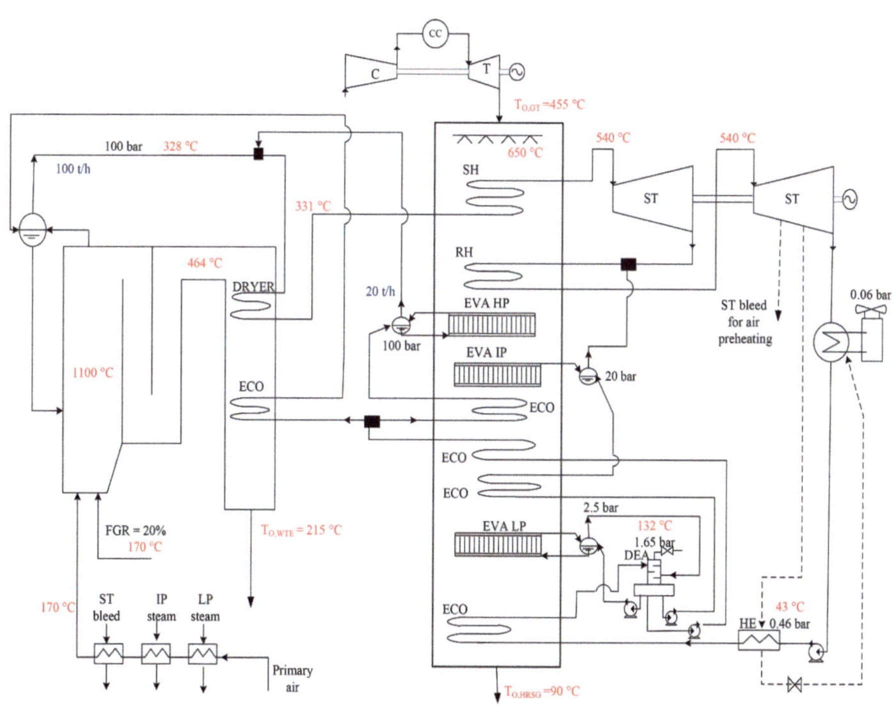

Fig. 5.5 Zabalgarbi integrated WTE–GT schematic layout. *C* compressor, *CC* combined cycle, *T* turbine, *SH* superheater, *ST* steam turbine, *RH* reheater, *EVA* evaporator, *HP* high pressure, *IP* intermediate pressure, *ECO* economizer, *LP* low pressure, *FGR* flue gas recirculation, *DEA* deaerator, *HE* general heat exchanger

Table 5.1 Zabalgarbi's technical data [16]

WTE section	
Type of waste	MSW and assimilated
Number of lines	1
Consumption of waste	33.08 t/h
Waste average lower heating value (LHV)	8000 kJ/kg
Productive capacity	240,000 t/year
Thermal power from waste	73.52 MW
Steam	
Saturated steam mass flow rate	100 t/h
Steam evaporative pressure (HP and LP)	100 and 25 bar
Steam-superheated and reheated temperature	540 °C
Natural gas	
Consumption of natural gas (GT+HRB)	13,870 N m^3/h
LHV of natural gas	38,992 kJ/N m^3
Thermal power introduced with natural gas	150 MW
Energy	
Gross electric power	99,176 MW
Internal consumption (with externals)	4,745 MW h (5,200 MW h)
Net power (grid connection) ST read	54,320 MW
Net power to the grid	94,431 MW
Net power (grid connection) GT read	40,111 MW
Net electricity generation	730,000 MW h/year
Efficiency and savings	
Total electrical gross efficiency	44.31 %
Overall net plant efficiency	42 %
Savings on conventional primary energy (thermal plant CC with natural gas comparison)	47 %
CO_2 emission avoided	300,000 t/year

WTE waste-to-energy, *LHV* lower heating value, *HP* high pressure, *LP* low pressure, *GT* gas turbine, *HRB* hybrid burner ring, *ST* steam turbine, *CC* combined cycle, *MSW* municipal solid waste

Main components in the system are: an MSW furnace with boiler for the generation of saturated steam, a GT generator, an HRSG working at 100 bar (equipped with auxiliary burners), an ST generator, a main condenser and an auxiliary one, and a flue gas cleaning system.

The saturated steam goes out of the main boiler (located in the WTE section) at 350 °C and 100 bar. To protect the boiler against corrosion, all exposed furnace tubes are coated with Inconel 625.

Saturated steam gets in the HRSG where it is heated up to 540 °C with heat coming from the exhaust of the GT and auxiliary burners. Superheated steam generated at 540 °C and 100 bar goes into the high-pressure ST to be expanded. Steam comes out of the high-pressure ST at a lower pressure (25 bar), goes again to the HRSG to be reheated at the same temperature (540 °C), and finally goes to the low-pressure ST. After condensation and before entering the furnace's boiler, the water is preheated with a low-pressure ST extraction.

The GT generator chosen was a General Electric LM6000 of 43 MW average gross capacity fueled with natural gas and equipped with an intake air cooling system (compression chiller). The GT exhaust temperature of the LM6000 is 455 °C; thus, to generate superheated steam at 540 °C, auxiliary burners (using natural gas and fresh air) are necessary, which work at temperatures around 650–700 °C.

LM6000 was chosen considering that, at the time of the project and authorizations, the plant was considered in a special Spanish regime (renewables and cogeneration), where the maximum size of the integrated plant should have not exceeded 100 MW; thus, considering WTE and consequently ST average capacity, the installed GT could not exceed 45 MW.

The effect of eliminating the superheated exchangers from the WTE section has been clearly stated: In conventional plants, maintenance (exchange of tubes) must be done every 1–2 years; in Zabalgarbi, the first rows of tubes were exchanged in the 6th year of operation.

Due to system integration, the power plant has a high flexibility. The following operations modes are feasible:

- Normal operation
- CC operation
- Fresh air mode
- Incineration and GT through bypass stack
- Incineration only
- Shutdown

Based on Zabalgarbi's operational and economic success, the Vizcaya Regional Government has focused its 2016 Integral Waste Management Plan on doubling the first plant's capacity. Investment costs for the WTE–GT integrated power plant have been estimated at about 136 million euros.

5.3.2 *Moerdijk WTE–GT Power Plant: The Dutch Solution*

The history of the WTE–GT power plant of Moerdijk, located in the south of the Netherlands, is closely related to the nearest MSW incinerator (property of the AZN company) and a CC plant (property of EPZ, the electricity-producing company for the Southern Netherlands). The two systems have been commissioned in the same period (1996 for the incinerator and 1997 for the CC power plant). Thus, the proximity and their contemporary construction are the main reasons for the idea of the integrated plant. For this reason, the CC has all the components and the design of an autonomous system. Figure 5.6 schematically presents the layout of the integrated plant, showing a complex three-pressure-level CC with reheating.

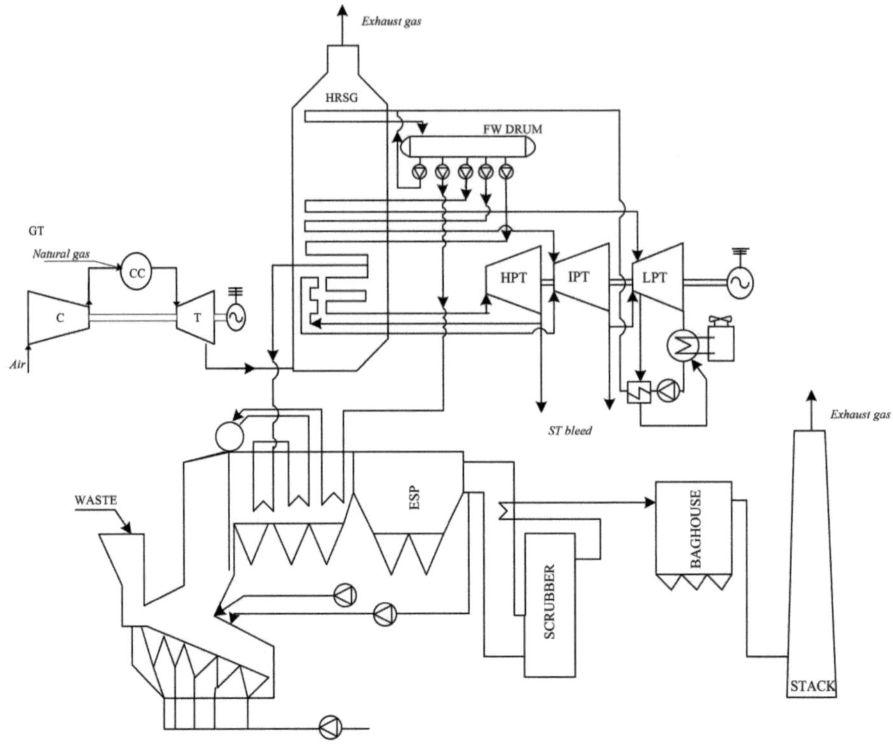

Fig. 5.6 Moerdijk integrated WTE–gas turbine *(GT)* schematic layout

The HRSG is used to superheat the steam coming from the furnace boiler to avoid corrosion problems in the WTE section. STs serve both the CC and WTE sections. The steam coming from the WTE boiler (at 100 bar) is mixed with the steam generated in the HRSG section. The superheated steam at 520 °C is then fed into the high-pressure ST. Middle-pressure and low-pressure steam are generated only in the HRSG section.

The incinerator has a capacity of about 80 t/h of waste with a lower heating value (LHV) of about 10,450 kJ/kg. Thus, the power introduced with waste is 232 MW_t. The CC is equipped with three GTs, each with an electric power of 60 MW. The total power of the integrated system is about 330 MW with an increase, with reference to separate production, of 18 MW.

5.3.3 *Takahama WTE–GT Power Plant: The Japanese Solution*

Near the city of Takahama in Japan, an MSW incineration plant was built in 1988, which is able to process 18 t/h of waste with an average calorific value of about 8400 kJ/kg. The primary objective of the incinerator is the disposal of waste, while

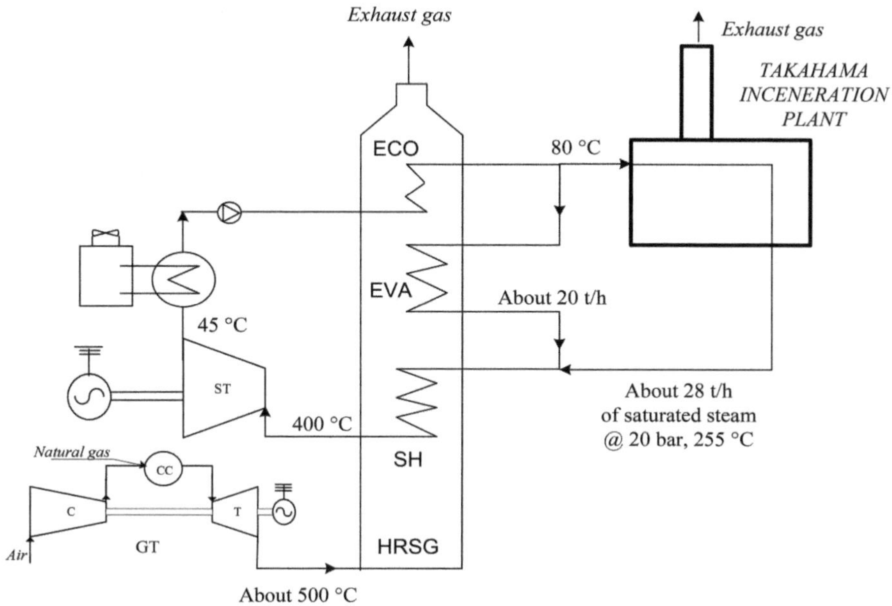

Fig. 5.7 Takahama integrated WTE–gas turbine *(GT)* schematic layout. *C* compressor, *CC* combined cycle, *T* turbine, *SH* superheater, *ST* steam turbine, *EVA* evaporator, *ECO* economizer, *HRSG* heat recovery steam generator

energy recovery was limited to the generation of only 1.3 MW, essentially equal to the auxiliary and internal consumptions of the plant.

Eight years later the idea of increasing the electricity production of the plant came up by integrating the WTE section with a small GT and a heat recovery boiler, where saturated steam generated in the incinerator at a temperature of 255 °C and a pressure of 20 bar could be superheated. In 1996 the integrated plant was realized: a 15-MW GT with a heat recovery boiler, a 10-MW ST, and the incinerator. The layout of the Takahama integrated power plant is shown in Fig. 5.7. The figure shows that water out of the ECO section is divided into two streams; thus, saturated steam (at 20 bar) is generated both in the WTE and HRSG section; after mixing, the steam is superheated in the heat recovery boiler at a temperature of 400 °C. The integrated plant has a total capacity of 25 MW.

References

1. Petrov MP, Martin AR, Hunyadi L (2002) Hybrid dual-fuel combined cycles: general performance analysis, ASME International Joint Power Generation Conference, Phoenix AZ, USA, June 2002
2. Consonni S (2000) Combined cycles for high performance, low cost, low environmental impact waste-to energy. Systems, Proceedings of ASME TURBOEXPO, May 8–11, 2000, Munich Germany

3. De Carli (2004) "FattibilitÃ tecnico – economica, inclusi gli aspetti autorizzativi, dell'integrazione inceneritore – ciclo combinato, con riferimento sia all'utilizzo di turbine a gas che di cicli a vapore". Technical report, CESI, Ricerca di Sistema (In italian)
4. William C, Stenzel PE, Dale M, Sopocy PE, Stanley E Pace, repowering existing fossil steam plants, SEPRIL. http://coal2nuclear.com/MSR%20-%20Repowering%20Existing%20Fossil%20Steam%20Plants%20-%20SEPRIL%20.pdf
5. Eber SM, Kettler DJ, Worsley CR, Fedele G (1989) Gas fired superheaters in a resource recovery installation. Proceedings of the American Power Conference, vol 51, Chicago, IL, April, pp 858–864
6. Lowry WL, Martin CA (1990) Economics of combined cycle waste-to-energy. Proceedings of the International Joint Power Generation Conference, Boston MA, USA, October, pp 31–37
7. Wiekmeijer IT (1990) Improvements in incinerators by means of gas turbine based cogen systems, ASME Paper 90-GT-180, Presented at the International Aeroengine Congress and Exposition, Brussels, Belgium, June
8. Terasawa H, Ogura M (1993) System to rationalize energy utilization at waste incineration plants. Proceedings of JSME—ASME International Conference on Power Engineering, Tokyo, September, vol 2, pp 591–594
9. Ito K, Yokoyama R, Shimoda M (1996) Optimal planning of a super waste incineration cogeneration plant. ASME Paper 96-GT-384, Proceedings of the International Gas Turbine and Aeroengine Congress and Exposition, Birmingham, UK, June
10. Sue M (1996) Performance characteristics of waste-to-energy system utilising steam-injected gas turbine. Transactions of the Japan Society of Mechanical Engineers, Part B, vol 62, Nr.597, May, pp 2013–2020
11. Otoma S, Mori Y, Terazono A, Aso T, Sameshima R (1997) Estimation of energy recovery and reduction of CO2 emissions in municipal solid waste power generation. Resour Conserv Recycl 20(2), 95–117
12. Holmgren R Hybridkraftverk för avfallseldning, M. Sc. Thesis, report Nr. 1998:228 CIV, ISSN: 1402-1617, ISRN: LTU-EX–98/228-SE, Division of Energy Technology, Luleå University of Technology, Luleå, Sweden, August 1998
13. Korobitsyn MA, Jellema P, Hirs GG (1999) Possibilities for gas turbine and waste incinerator integration. Energy 24(9):783–793
14. Udomsri S, Martin AR, Fransson TH.(2010) Economic assessment and energy model scenarios of municipal solid waste incineration and gas turbine hybrid dual-fueled cycles in Thailand. Waste Manag 30:1414–1422
15. Petrov MP, Hunyadi L (2002) Municipal solid waste boiler and gas turbine hybrid combined cycles performance analysis. 1st International Conference on Sustainable Energy Technologies 12–14 June, Porto, Portugal paper n. EES6
16. CNIM and SENER (2005) Bizkaia waste to energy plant project, summary report. Contract No BM 80/96 ES-PO
17. SENER Ingeniería y Sistemas S.A., High efficiency technologies applied for waste to energy (WTE)
18. AZN and EPZ, Combination of a waste incineration plant and a combined cycle power plant, project No BM/349/93/NL
19. Japanese National Team (1997) First Japanese waste incinerator and combined cycle power plant repowered with a gas turbine. Caddet Energy Effciency, Waste and Biogas, Newsletter No 2, 1997

Chapter 6
WTE–GT Steam/Waterside Integration: Thermodynamic Analysis on One Pressure Level

This chapter focuses on waste-to-energy (WTE) and gas turbine (GT) integrated configurations concerning one-pressure-level heat recovery steam generator (HRSG). The thermodynamic and parametric analysis of steam/waterside-integrated WTE–GT power plant has been carried out, first of all, with the aim to investigate the logic governing plants and that it should match in terms of steam production as a function of the thermal power generated. Steam generation, optimum plant match condition, inlet and outlet conditions of heat exchangers, etc., as a consequence of system integration are analyzed and explained. A sensitivity analysis, varying with evaporative pressure and HRSG inlet conditions, has also been presented to investigate the influence of operative parameters on steam mass flow rate.

Starting with a simple integrated plant layout, several configurations have been proposed and analyzed. Positive aspects and limitations of each layout have been investigated and discussed.

The following paragraphs assess and define, for a given layout and operative conditions, the optimum WTE–GT plant match in terms of system input of thermal power to maximize steam generation and plant performance and to minimize discharged outlet temperature.

6.1 Thermodynamic Analysis of Steam Production

Focusing on a waste-to-energy–gas turbine (WTE–GT) steam/waterside-integrated power plant, the simplest layout with one-pressure-level HRSG is shown in Fig. 6.1 (named Layout_1). This configuration can be regarded as a starting point to investigate the logic concerning steam production in an integrated plant.

In the proposed layout, while saturated steam gets produced in the waste boiler, water gets preheated and steam superheated in the HRSG economizer (ECO) and superheater (SH), respectively; one bleed from the steam turbine (ST) is present to preheat the primary air necessary for waste combustion and to feed the deaerator (DEA).

L. Branchini, *Waste-to-Energy*, DOI 10.1007/978-3-319-13608-0_6

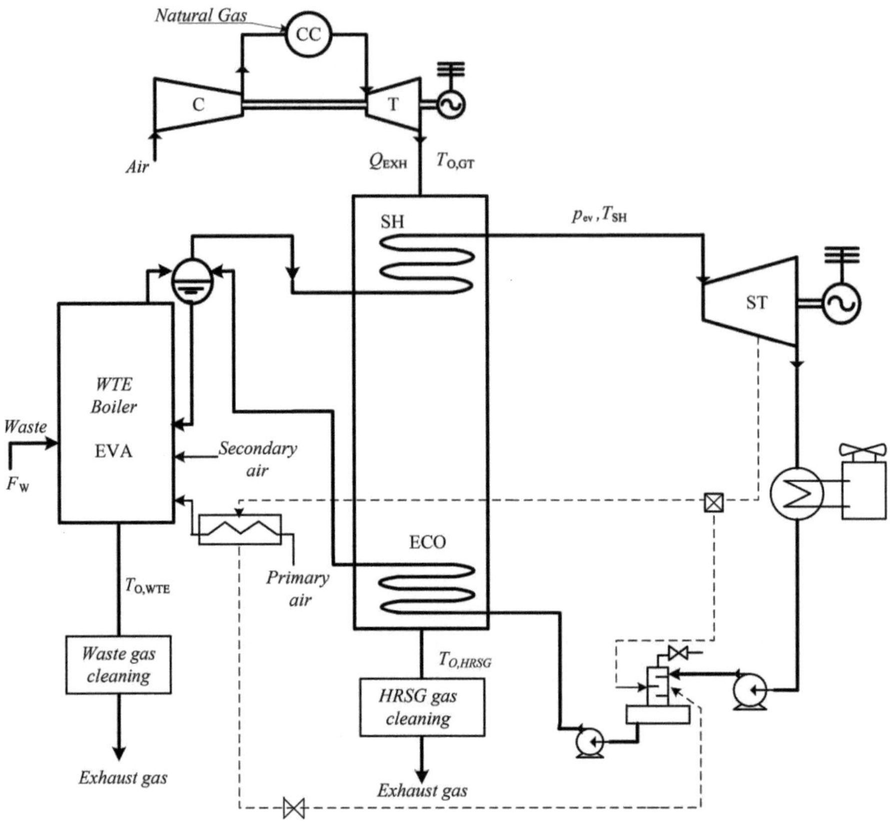

Fig. 6.1 WTE–GT integrated power plant layout (Layout_1). *C* compressor, *CC* combined cycle, *T* turbine, *SH* superheater, *ST* steam turbine, *EVA* evaporator, *ECO* economizer, *HRSG* heat recovery steam generator

As we focus on steam production, we find that steam generation is affected by:

- The thermal power discharged from the GT entering the HRSG section:

$$Q_{\mathrm{EXH}} = \dot{m}_{\mathrm{EXH}} c_{p,\mathrm{EXH}} \left(T_{\mathrm{O,GT}} - T_{\mathrm{ref}}\right), \tag{6.1}$$

where $T_{\mathrm{O,GT}}$ is the GT outlet temperature, T_{ref} is the reference temperature assumed to be equal to 15 °C, and $c_{p,\mathrm{EXH}}$ is the specific heat of the exhaust gases, a function of $T_{\mathrm{O,GT}}$.

- The input power introduced with waste in the WTE boiler:

$$F_{\mathrm{W}} = \dot{m}_{\mathrm{W}} LHV_{\mathrm{W}}, \tag{6.2}$$

where $\dot{m}_{\mathrm{W}}$ and lower heating value (LHV_{W}) are waste mass flow rate and lower heating value, respectively.

Consequently, energy balance allows us to evaluate the superheated steam produced in the HRSG ($\dot{m}_{s,HRSG}$) and the saturated steam generated by the WTE boiler ($\dot{m}_{s,WTE}$) with the following equations:

$$\dot{m}_{s,HRSG} = \frac{\xi Q_{EXH}}{\Delta h_{SH} + \Delta h_{ECO}}, \tag{6.3}$$

$$\dot{m}_{s,WTE} = \frac{\eta_{boil} F_W}{\Delta h_{lat} + \Delta h_{sc}}, \tag{6.4}$$

where:

- ξ is the HRSG effectiveness, which mainly depends on the HRSG inlet and outlet temperatures, $T_{O,GT}$ and $T_{O,HRSG}$.
- η_{boil} is the WTE boiler efficiency, which is mainly due to outlet temperature of WTE gases, $T_{O,WTE}$, and the ratio of air to fuel waste mass flow rates.
- Δh_{SH} and Δh_{ECO} are steam- and water-specific enthalpy, which increases in the SH and in the ECO, respectively.
- $\Delta h_{lat} + \Delta h_{sc}$ is the specific enthalpy rise in the evaporator (EVA), which is the sum of latent heat and subcooling enthalpy. Once selected, the evaporative pressure value (p_{ev}), Δh_{lat} is constant, while Δh_{sc} is equal to the difference between evaporative and ECO outlet temperature. Thus, it can be equal to or higher than a minimum value imposed to avoid water evaporation in the ECO section depending on the Q_{EXH} value, which will be detailed later.

The adopted layout proposes to have the same value of steam mass flow rate in the HRSG and in the WTE sections. Consequently, by equating Eqs. (6.3) and (6.4) the following must be satisfied:

$$\frac{Q_{EXH}}{F_W} = \frac{\eta_{boil}}{\xi} \frac{\Delta h_{SH} + \Delta h_{ECO}}{\Delta h_{lat} + \Delta h_{sc}}. \tag{6.5}$$

Figure 6.3 shows how Eq. (6.5) could be solved qualitatively to calculate steam production as a function of GT-discharged thermal power (Q_{EXH}), assuming the values $T_{O,GT}$, p_{ev}, and F_W remain constant.

In particular, $\dot{m}_{s,max}$, calculated from Eq. (6.4), represents the upper limit of steam production, regardless of the HRSG size, determined only by the thermal power introduced with waste, F_W.

The main design point (point B in Fig. 6.2) can be identified, corresponding to optimum GT-discharged thermal power, Q^*_{EXH}.

For GT-discharged thermal power higher than Q^*_{EXH} (point B) and moving from B to A, the following takes place:

- HRSG outlet temperature ($T_{O,HRSG}$) increases (decreasing HRSG effectiveness, ξ); thus, it overcomes its minimum value (Fig. 6.3a).

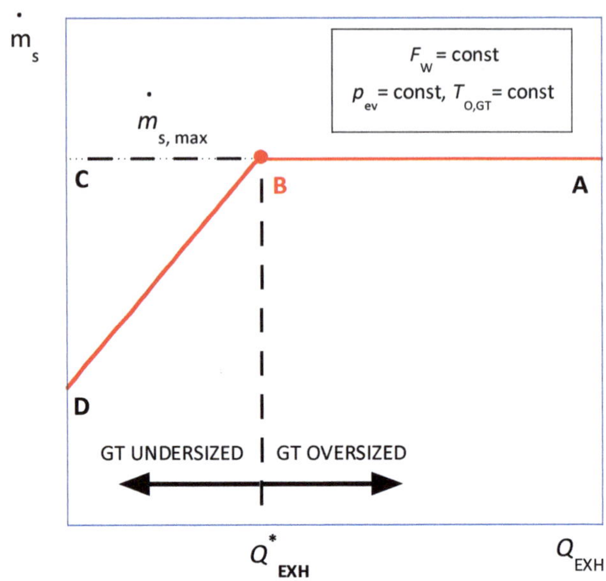

Fig. 6.2 Steam to waste mass flow rate as a function of gas turbine *(GT)*- discharged thermal power

- No change in subcooling temperature difference, ΔT_{sc} (Fig. 6.4b), which keeps its minimum value.
- Superheated steam temperature, T_{SH} (Fig. 6.3c), remains constant and at its maximum value if we assume a minimum approach difference between $T_{O,GT}$ and T_{SH}.

A reduction of Q_{EXH} under Q^*_{EXH} can be managed with two possible strategies:

1. Dotted line B → C: To keep the steam mass flow rate at its maximum value $\dot{m}_{s,max}$, a reduction in T_{SH} is required. In this case, the thermal power discharged from the GT is not enough to superheat, at maximum level, the steam that the WTE boiler can produce.
2. Continuous line B → D: T_{SH} can be kept constant so that the efficiency of the steam cycle is not compromised, thus reducing the steam mass flow rate and increasing ΔT_{sc}.

In this case, the difference between the EVA and ECO outlet temperatures is higher than the minimum value allowed; this increase in the subcooling temperature is equivalent to the introduction of an ECO section in the WTE boiler.[1]

Therefore, point B represents the best choice (maximum T_{SH}, minimum ΔT_{sc}, and $T_{O,HRSG}$) for the assumed layout.

[1] If an ECO section was placed into the WTE boiler, point D will show an increase in steam mass flow rate, due to a lower temperature water entering the WTE boiler, causing a further reduction of $T_{O,WTE}$.

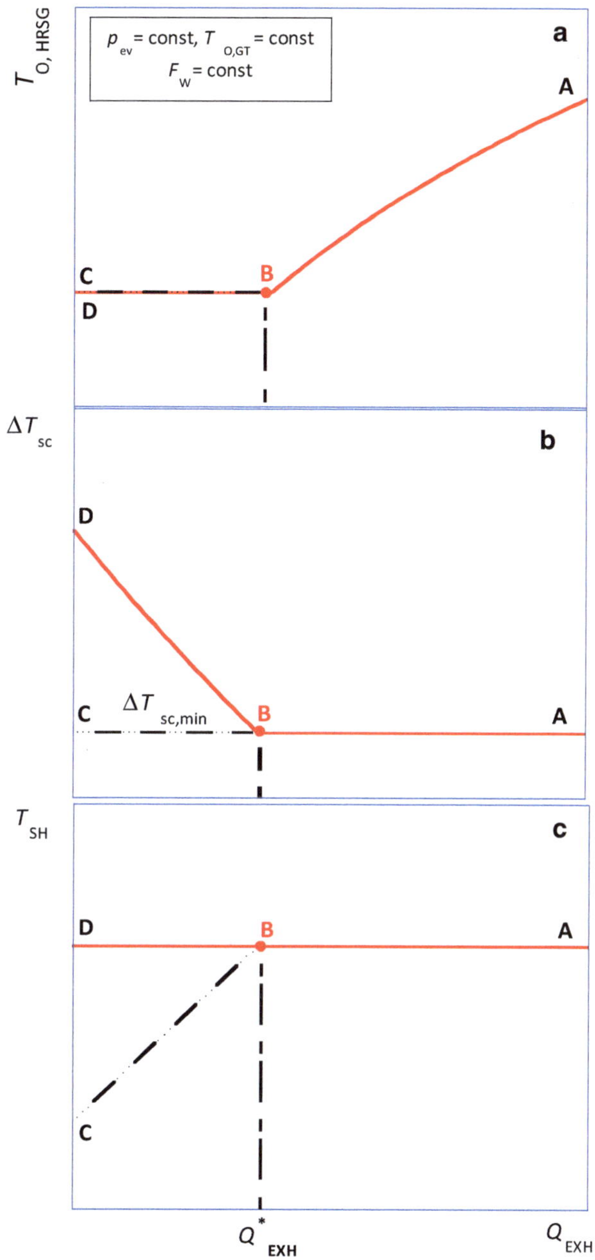

Fig. 6.3 $T_{O,HRSG}$ (**a**), ΔT_{sc} (**b**), and T_{SH} (**c**) realized as functions of GT-discharged thermal power

It is possible to operate the plants by keeping the thermodynamic parameters at their optimum values (maximum T_{SH} and minimum $T_{O,HRSG}$, line B–D), but, in this case, steam mass flow rate is lower than its maximum value. Thus, the WTE section turns out to be undersized.

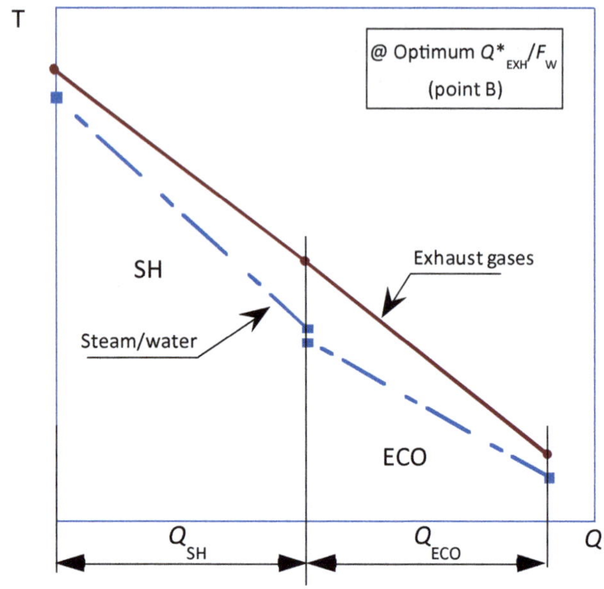

Fig. 6.4 *T–Q* diagram for the HRSG section of the integrated plant. *SH* superheater, *ECO* economizer

However, it is not convenient to operate by keeping the steam mass flow rate $\dot{m}_{s,\max}$ at the maximum, but with Q_{EXH} far from Q^*_{EXH} (line A–B or line B–C); this means that one has to operate with low HRSG effectiveness (A–B) or with low thermodynamic steam cycle efficiency (B–C).

This analysis highlights that, for a given evaporative pressure and GT outlet temperature, an optimum plant match, in terms of thermal power ratio, Q^*_{EXH}/F_W, is found.

Beyond this value, an increase in HRSG thermal power input does not lead to additional benefits in terms of generated steam mass flow or HRSG effectiveness.

Figure 6.4 shows the *T–Q* diagram for the HRSG section, corresponding to the optimum value (point B, Q^*_{EXH}/F_W). As previously mentioned, one of the main advantages of a WTE–GT integrated plant is the reduction of the HRSG irreversibility caused by high mean temperature differences. As stressed in the figure, removing the EVA from the HRSG reduces the distance between exhaust gases and steam/water lines, with respect to a typical one-pressure-level HRSG *T–Q* diagram.

Finally, for the investigated layout, it must be pointed out that $T_{O,WTE}$ strongly depends on evaporative temperature; thus, the choice of evaporative pressure mainly affects η_{boil}; the higher the p_{ev}, the lower is the WTE boiler efficiency.

6.1.1 *Influence of Evaporative Pressure and GT Outlet Temperature on Steam Production*

In order to investigate the influence of steam cycle parameters on optimum WTE–GT plant match, a parametric analysis has been performed by varying steam evaporative pressure and GT outlet temperature.

Figures 6.5 and 6.6 show, qualitatively, the influence of three different evaporative pressures ($p_{ev} < p'_{ev} < p''_{ev}$) and $T_{O,GT}$ values ($T_{O,GT1} < T_{O,GT2} < T_{O,GT3}$) on steam mass flow rate production, respectively. Focusing on Fig. 6.5, an increase in the evaporative pressure value and a decrease in Δh_{lat} can lead to an increase in steam mass flow rate as expressed in Eq. (6.4).

If the thermal power introduced in the GT exhaust is not enough to bring about maximum steam flow rate (points B, B′, and B″), the amount of produced steam is lower than the maximum value, and its variation is not dependent on p_{ev}. By

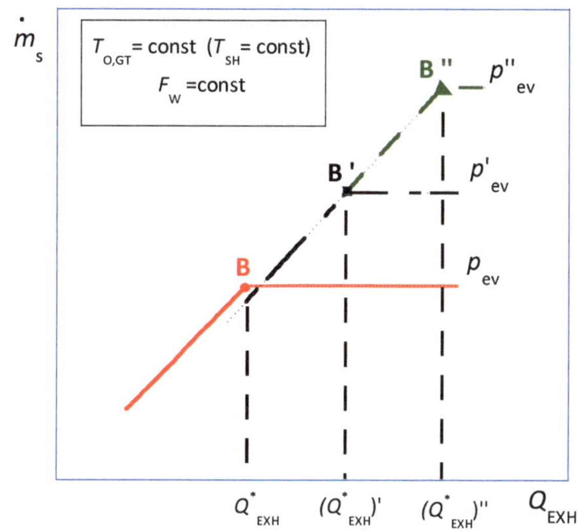

Fig. 6.5 Steam mass flow rate as a function of GT-discharged thermal power for different evaporative pressure values

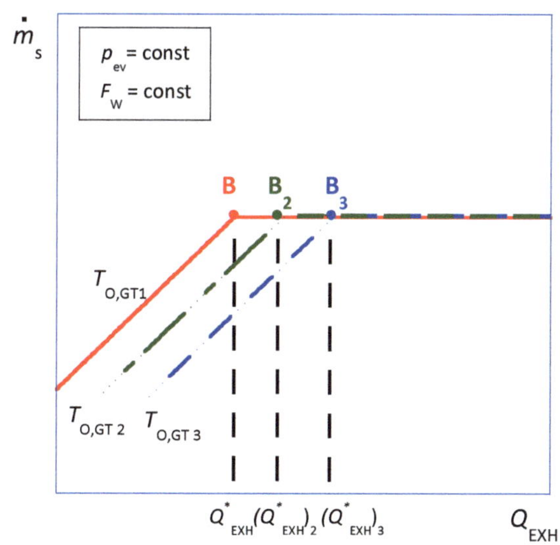

Fig. 6.6 Steam mass flow rate as function of GT-discharged thermal power for different GT outlet temperature

increasing Q_{EXH} over Q^*_{EXH}, a constant trend is achieved for every evaporative pressure value. Moreover, by increasing the evaporative pressure, an optimum condition (B, B′, B″, respectively, for p_{ev}, p'_{ev}, and p''_{ev}) is achieved, which is characterized by a greater optimum thermal power ratio, that is, the Q^*_{EXH} / F_W value.

The influence of $T_{O,GT}$ (and consequently T_{SH}) is shown in Fig. 6.6; for Q_{EXH} below the optimum point, an increase in GT outlet temperature (and on steam-superheated one) decreases the steam mass flow rate. Once the optimum point is reached (B, B_2, B_3, respectively, for $T_{O,GT1}$, $T_{O,GT2}$, and $T_{O,GT3}$), steam production is the only function of the input waste thermal power and is not affected by $T_{O,GT}$.

6.2 Numerical Results

The main results of the study are collected and shown in Fig. 6.7. A constant waste composition and LHV_W value have been assumed, as reported in Table 6.1. In Table 6.2, additional assumptions used for numerical analysis are listed.

The grid of Fig. 6.7 represents optimum plant match (point B) for different evaporative pressure and GT outlet temperature values.

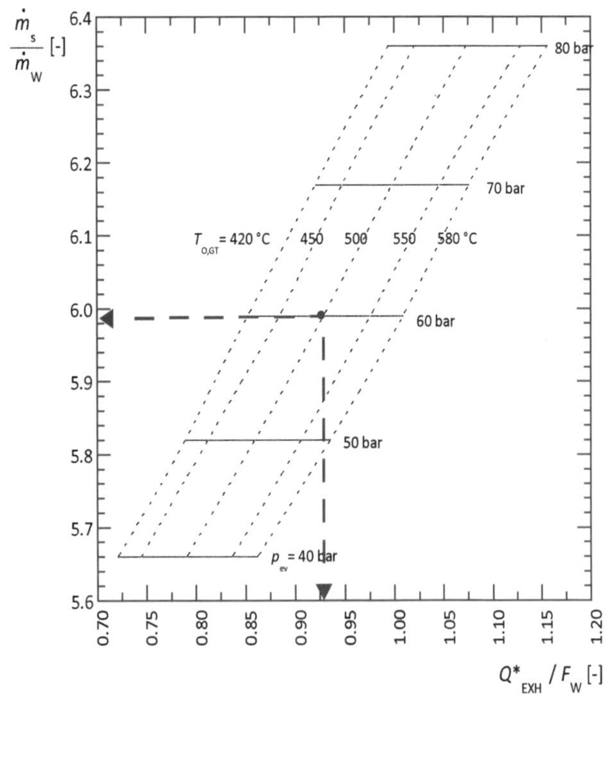

Fig. 6.7 Grid representing optimum plant match for a given evaporative pressure and GT outlet temperature as a function of thermal power input ratio

Table 6.1 Municipal solid waste (MSW) composition and lower heating value (LHV) assumed for the analysis

Proximate analysis (% by weight)	Humidity	23.0	% of volatile matter	C	52.5
	Ashes	15.0		H	7.5
	Volatile matter	62.0		O	38.5
				N	1.3
				S	0.2

LHV_W = 11.85 MJ/kg$_W$ (2.83 kcal/kg$_W$)

Table 6.2 Main assumptions

ST extraction pressure for primary air heating and deaerator (bar)	3.5
ST isentropic efficiency (–)	0.85
ST outlet quality (–)	0.84 ÷ 0.95
Condenser pressure (bar)	0.1
Deaerator working pressure (bar)	3
Oxygen in exhaust dry gases (vol. %)	7
WTE exhaust gas recirculation (vol. %)	15
WTE exhaust gas recirculation temperature (°C)	150
Primary air temperature (°C)	130
Secondary air temperature (°C)	50
ΔT pinch point (°C)	10
ΔT approach (°C)	20
ΔT_{sc} (°C)	10
$T_{O,WTE}$ (°C)	≥160
$T_{O,HRSG}$ (°C)	≥110

ST steam turbine, *WTE* waste-to-energy

Intersecting the line at constant pressure with the line at constant temperature, the steam mass flow rate and the optimum ratio of HRSG to WTE thermal power input is obtained.

For example, at evaporative pressure equal to 60 bar and a GT outlet temperature of 500 °C (corresponding to 480 °C of superheated steam temperature), about 6 kg/s of steam mass flow rate is generated for every kg/s of waste. Moreover, the optimum plant match corresponding to $Q^*_{EXH}/F_W = 0.93$ is found.

As highlighted in Fig. 6.7, high evaporative pressure and superheated steam temperature means high Q^*_{EXH}/F_W. Thus, the higher the steam cycle parameters, the higher must be the GT-discharged thermal power per unit of thermal power introduced with waste.

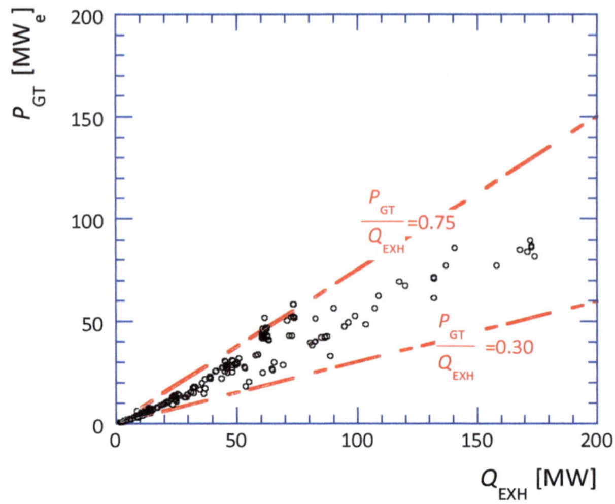

Fig. 6.8 GT power output as a function of GT-discharged heat for commercial GT units

6.2.1 Optimum Plant Match in Terms of Electric Power Ratio

Optimum plant match can be expressed in terms of electric power output rather than the thermal power input.

In order to evaluate the GT electric size (P_{GT}) that must be chosen to realize the integrated power plant of Fig. 6.1, the following equation can be formulated:

$$\frac{P_{GT}}{P_{WTE}} = \frac{P_{GT}}{Q_{EXH}} \cdot \frac{Q_{EXH}}{F_W} \cdot \frac{F_W}{P_{WTE}}, \tag{6.6}$$

where P_{WTE} is the electric capacity of a traditional WTE power plant fed with the same waste input as that of the integrated system.

Taking into account that

- the electric capacity of a GT is a fraction of the discharged thermal power Q_{EXH} ranging from 0.30 to 0.75 depending on the GT characteristics (see Fig. 6.8) and that
- for a traditional WTE power plant, the efficiency P_{WTE}/F_W typically ranges from 0.25 to 0.30.

From Eq. (6.6) it can be concluded that optimum matching requires one to select a GT with an electric power output ranging from the same WTE size (minimum P_{GT}/Q_{EXH} and maximum P_{WTE}/F_W) up to three times of the WTE original plant (maximum P_{GT}/Q_{EXH} and minimum P_{WTE}/F_W).

To keep the GT size at its minimum, the P_{GT}/Q_{EXH} ratio has to be the lowest. Since a relationship between P_{GT}/Q_{EXH} and the GT efficiency (η_{GT}) can be identified as

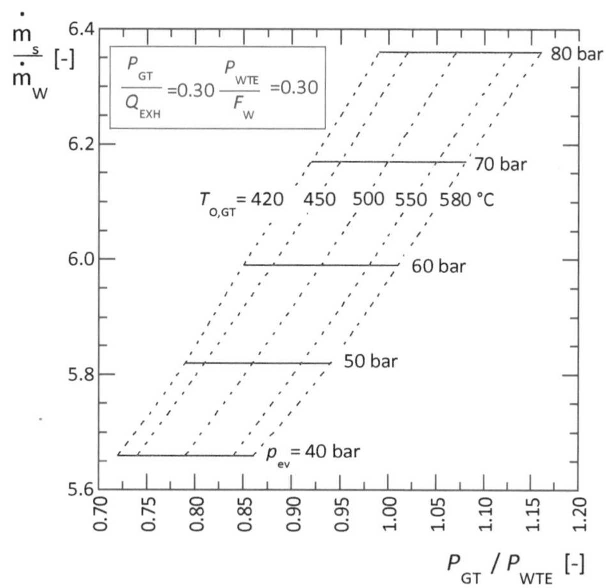

Fig. 6.9 Grid representing optimum plant match as a function of output power ratio for WTE efficiency of 30 % and a ratio of GT electric power output to the discharged heat, which is equal to 0.3

$$\eta_{GT} \approx \frac{1}{1+\left(\frac{Q_{EXH}}{P_{GT}}\right)}, \tag{6.7}$$

it follows that the GT with low efficiency is the best choice for this purpose.

Figure 6.9 shows, as an example, the steam-to-waste mass flow rate ratio as a function of electric power output P_{GT}/P_{WTE}. The figure refers to WTE efficiency of 30 % and a ratio of GT electric power output and discharged heat equal to 0.30. Thus, optimum conditions expressed in terms of the thermal power ratio (see Fig. 6.7) agree with those found in terms of power output ratio $\left(\frac{P_{GT}}{P_{WTE}} = \frac{Q^*_{EXH}}{F_W}\right)$. However, Fig. 6.10 shows that optimum values in terms of electric power ratio are about three times those found in Fig. 6.9, assuming WTE efficiency of 25 % and a ratio of GT electric power output to discharged heat equal to 0.75.

6.2.2 Traditional WTE Versus Integrated Plant: Steam Turbine Capacity

The analysis compared the steam mass flow rate produced by an integrated plant with that generated by a traditional WTE fed with the same amount of waste power input and highlighted how a considerable increase can be achieved.

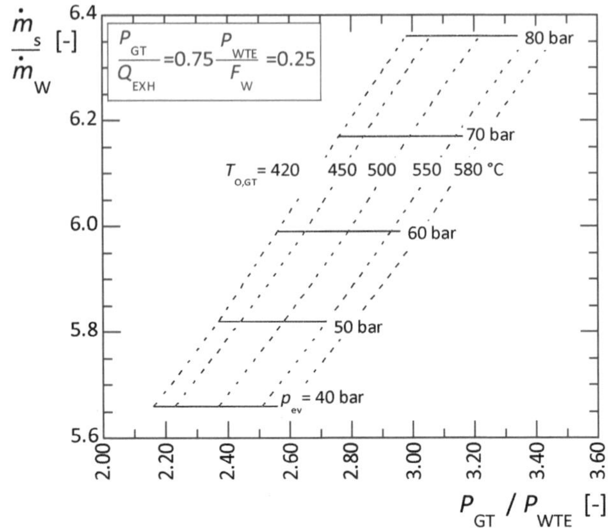

Fig. 6.10 Grid representing optimum plant match as a function of output power ratio for WTE efficiency of 25 % and a ratio of GT power output to discharged heat equal to 0.75

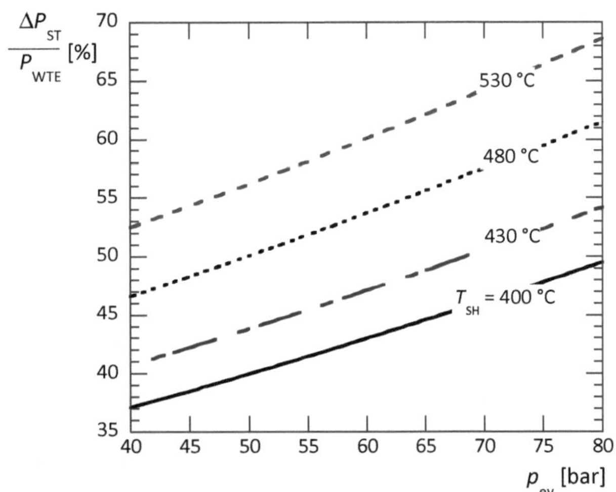

Fig. 6.11 Increase in steam turbine power output as a function of evaporative pressure and GT outlet temperature

As a consequence, the capacity of the ST in the WTE–GT integrated system will be higher than the one employed in a WTE stand-alone power plant. It can be expressed as

$$P_{ST} = P_{WTE} + \Delta P_{ST}, \tag{6.8}$$

where ΔP_{ST} is proportional to the steam mass flow increase.

Figure 6.11 shows the percentage increment of ST power output with respect to a WTE stand-alone power plant and highlights that it ranges, for the considered cases, from 35 to 70 %. It reaches its highest value for the maximum evaporative pressure and steam-superheated temperature.

6.3 Concluding Remarks on Thermodynamic Analysis

The study suggests that an optimum WTE–GT plant match in terms of the system's thermal power input must be pursued to maximize steam generation and steam-superheated temperature and to minimize exhaust gas temperature.

Deviation from optimum conditions can be resolved through the following steps: (i) oversize the GT without additional benefits in terms of generated steam mass flow rate or the effectiveness of HRSG, (ii) depress the WTE section, decreasing the amount of generated steam, or (iii) work with low steam cycle thermodynamic efficiency. Thus, the thermodynamic and parametric analysis provide useful guidelines in selecting an optimum GT size to match WTE–GT maximum performance. The study highlights that the higher the steam cycle parameters (evaporative pressure and steam-superheated temperature), the higher must be the GT-discharged thermal power with respect to the thermal power introduced with waste.

A correspondence between optimum thermal and electric power ratio can be achieved only if the GT efficiency is low, otherwise, optimum values in terms of electric power ratio can be up to three times that found corresponding to input thermal power.

Comparing steam mass flow rate produced in the WTE–GT integrated system with a WTE stand-alone, we find significant increases. Moreover, percentage gains in steam mass flow rate increase the increasing evaporative pressure and steam-superheated temperature. Increments in ST power output proportional to steam mass flow rate increase, and new ST capacity, for the integrated plant, has been derived: Power output increase ranges from 35 to 70 %, reaching its highest values for the maximum evaporative pressure and superheated temperature.

6.4 WTE–GT Proposed Layouts for a One-Pressure-Level HRSG

In this paragraph, WTE–GT layouts for a one-pressure-level HRSG are proposed and investigated. Simulations have been performed using Gatecycle software. Integrated system layouts are optimized, case after case, in order to exploit both the thermal power input with the waste and the GT-discharged thermal power. For each one of the proposed configurations, steam mass flow generation is analyzed and discussed as a function of GT-discharged thermal power, and the optimum plant match condition is identified. Based on common assumptions, the numerical results for each one of the WTE–GT integrated plant layouts are highlighted.

Municipal solid waste (MSW) composition and, as a consequence, waste LHV assumed for the analysis have been taken according to Table 6.1.

Table 6.3 Assumption of the composition of GT exhaust gases [2]

Molecular weight	28.6344
Molecular fraction composition	
Oxygen, O_2	0.1531
Nitrogen, N_2	0.7610
Water, H_2O	0.0510
Carbon monoxide, CO	0.0000
Carbon dioxide, CO_2	0.0258
Methane, CH_4	0.0000
Hydrogen, H_2	0.0000
Argon, Ar	0.0091
Carbonyl sulphide, COS	0.0000
Hydrogen sulphide, H_2S	0.0000
Sulphur dioxide, SO_2	0.0000
Specific enthalpy value	
h_{EXH} (at 500 °C) (kJ/kg)	519.78
h_{rif} (at 15 °C) (kJ/kg)	−57.04

Table 6.4 WTE outlet temperature as a function of evaporative pressure

p_{ev} (bar)	$T_{O,WTE}$ (°C)
40	260
50	274
80	305

Being interested in evaluating the optimum plant match condition in terms of thermal power discharged by a GT exhaust, we have assumed a typical GT exhaust gas composition according to [2] and as summarized in Table 6.3. The reference composition entering the HRSG section, shown in Table 6.4, refers to an air-to-fuel (CH_4) mass flow rate ratio equal to 68.94. Taking into account the stoichiometric ratio of combustion of air to methane (17.23), the excess air in the composition of exhaust gases turns out to be equal to 4, in line with typical GT combustion values.

The remaining steam cycle common assumptions are reported in Table 6.2.

The following aspects have been considered in the analysis of factors that affect waste combustion:

- The total air required for the waste boiler is set to obtain a prescribed oxygen concentration in dry exhaust gases; based on the oxygen content in the exhaust gas, the program automatically calculates the excess of air with respect to a stoichiometric combustion. For the assumed value of oxygen content in dry exhaust gases, the combustion air-to-waste mass flow ratio turns out to be equal to 6.55 (thus, the excess air fraction is equal to 0.63).
- Moreover, in the program, the primary air is specified in relation to the mass flow rate of the fuel, that is, as a mass fraction (mass flow of primary air/mass flow of fuel) assumed equal to 4.1; thus, 63 % of the total amount of combustion

air is identified as primary air, while the remaining part is secondary combustion air.

- Primary and secondary combustion air are supposed to be preheated up to a temperature of 50 °C by exploiting grate cooling.
- The integrated boiler (representing three vertical radiation passes, see Chap. 3, Sect. 3.1.3) has been modeled taking into account the contribution of both the thermal radiation and convective heat transfer; the overall heat transferred to EVAs and the waterwalls bordering the combustion chamber are mainly due to thermal radiation (about 87 % of the total heat exchanged).
- The assumed MSW composition, the calculated combustion air mass flow, the assumed exhaust flue gas recirculation (FGR), and combustion parameters lead to an adiabatic combustion temperature equal to about 1185 °C, which is in line with typical values for a WTE integrated boiler.
- A convective heat exchanger section (convective pass, see Chap. 3, Sect. 3.1.3) has also been modeled. The temperature at the convective section inlet has been fixed according to a typical WTE value, equal to 650 °C (Chap. 3, Sect. 3.1.3).
- The effectiveness of the heat exchanger has been kept at an upper limit to restrict heat exchanger surfaces.

Numerical results for WTE–GT integrated layout (Layout_1 in Fig. 6.1) are shown in Figs. 6.12 and 6.13 for *T–Q* diagram for the HRSG section corresponding to the

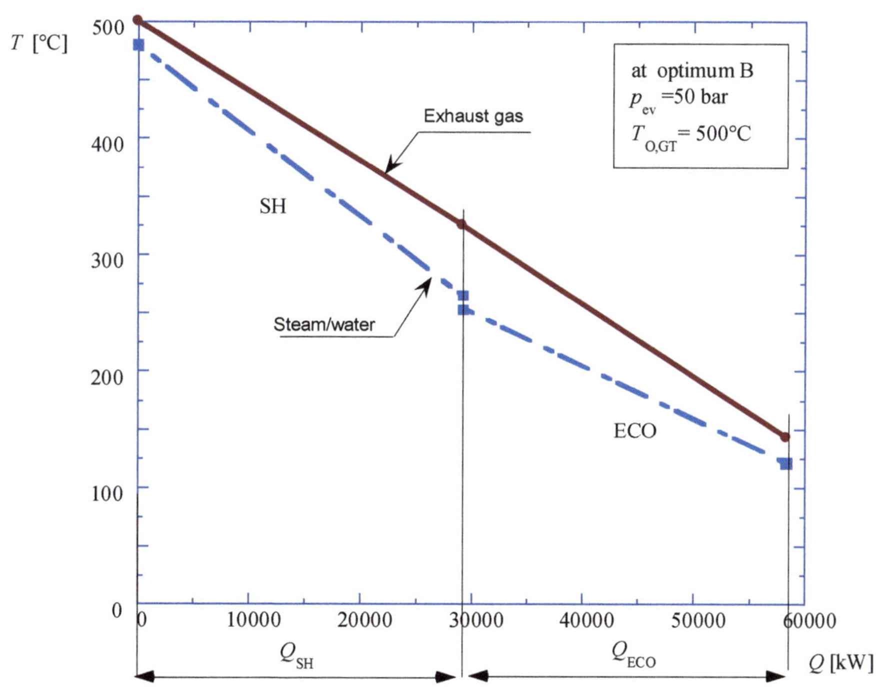

Fig. 6.12 *T–Q* diagram for HRSG section for Layout_1. *ECO* economizer

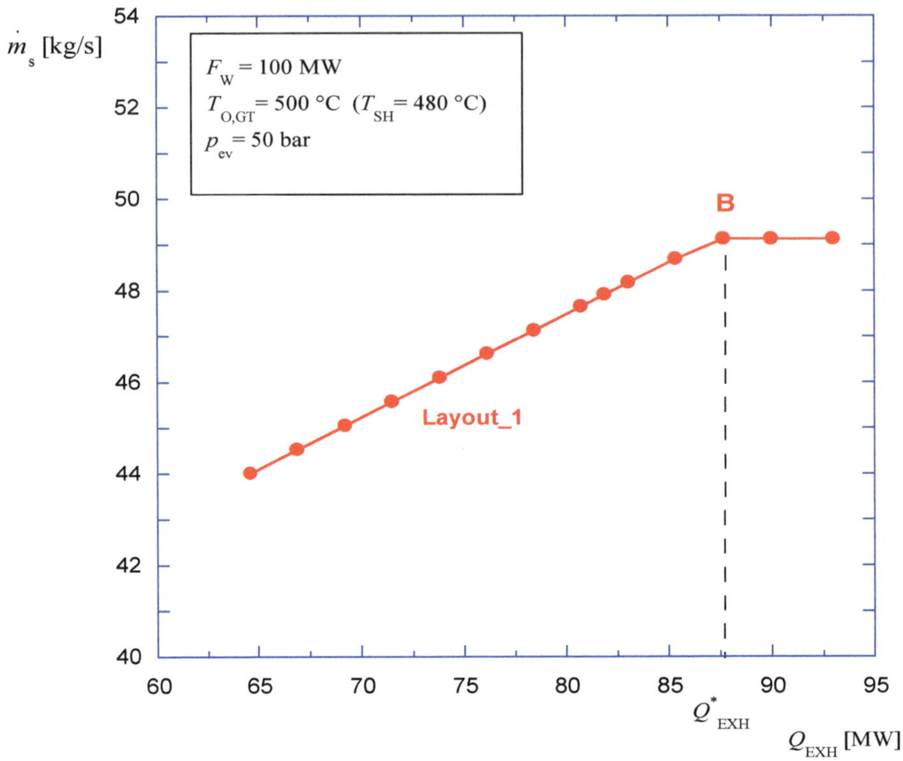

Fig. 6.13 Steam mass flow rate for Layout_1 as a function of GT-discharged thermal power

optimum point and steam mass flow rate generated as a function of GT-discharged thermal power, respectively.

Numerical results confirm the thermodynamic analysis described in the previous paragraph: The steam mass flow rate increases linearly, reaching the maximum value corresponding to the optimum plant match (point B), above which a constant trend is reached. Optimum plant match condition, for the considered layout, turns out to be equal to about 87.7 MW corresponding to maximum steam mass flow rate of 49.12 kg/s. The lowest considered value (corresponding to a steam mass flow rate equal to about 44 kg/s) agrees with the highest effectiveness assumed for SH heat exchanger. By contrast, there is no upper limit, but the constant trend has been interrupted in proximity to point B because, as previously described, it is not convenient to operate with values higher than Q^*_{EXH}.

For the investigated layout, it must be noticed that $T_{O,WTE}$ strongly depends on the evaporative temperature; having assumed only the presence of evaporative heat exchanger in the WTE integrated boiler, the exhaust gas temperature is a function only of the evaporative pressure and the pinch point assumed, as reported in Table 6.4.

Table 6.5 Layout_1 main results

Layout_1	
F_W (MW)	100
p_{ev} (bar)	50
$T_{O,GT}$ (°C)	500
T_{SH} (°C)	480
$\dot{m}_{EXH}$ (kg/s)	152
Q^*_{EXH} (MW)	87.68
Q^*_{EXH}/F_W (–)	0.88
$\dot{m}_{s,max}$ (kg/s)	49.12
$T_{O,HRSG}$ (°C)	143
$T_{O,WTE}$ (°C)	274
P_{ST} (MW)	45.76

The main results of the proposed layout, corresponding to optimum plant match condition (point B), are summarized in Table 6.5. The optimum Q^*_{EXH}/F_W corresponding to an evaporative pressure of 50 bar is below the unit (Q^*_{EXH}/F_W equal to 0.88). This means that the thermal power introduced with waste is greater than that introduced with the GT exhaust gases. Thus, the biggest supply in terms of thermal power introduced is assigned to the WTE section.

The main negative aspect of the proposed layout is the high WTE exhaust gas temperature: A relatively high amount of heat is discharged from the WTE boiler ($T_{O,WTE}$ higher than 270 °C), which is not completely used at the minimum temperature allowed (160 °C).

Layout_2 and Layout_2bis The next proposed layout (Layout_2) is schematically shown in Fig. 6.14. As can be seen in the figure, one ECO section is present in the WTE and one in the HRSG section. Water out of the DEA is first fed to the ECO1, which is placed in the WTE convective pass, where part of the economization is performed; economization is completed in the ECO2, inside the HRSG section.

The idea to share water economization comes from results of the Layout_1 that tried to exploit the thermal power available in the boiler, minimizing the WTE outlet temperature. Compared with the previous layout, there was considerable reduction of Q^*_{EXH} corresponding to the optimum point (point B″, Fig. 6.15). Optimum plant match condition is found corresponding to $\left(Q^*_{EXH}\right)'$ and equal to 81.9 MW. A decrease in Q^*_{EXH} equal to 6.6% of Layout_1 was obtained, while the maximum amount of steam produced remains the same. Focusing on steam production, for a fixed evaporative pressure and steam-superheated temperature, Layout_2 shows an increase in steam mass flow rate compared to Layout_1 for Q_{EXH} lower than the optimum value (see Fig. 6.15). The increase in steam mass flow can be explained considering the presence of the first ECO section in the WTE—allowing an

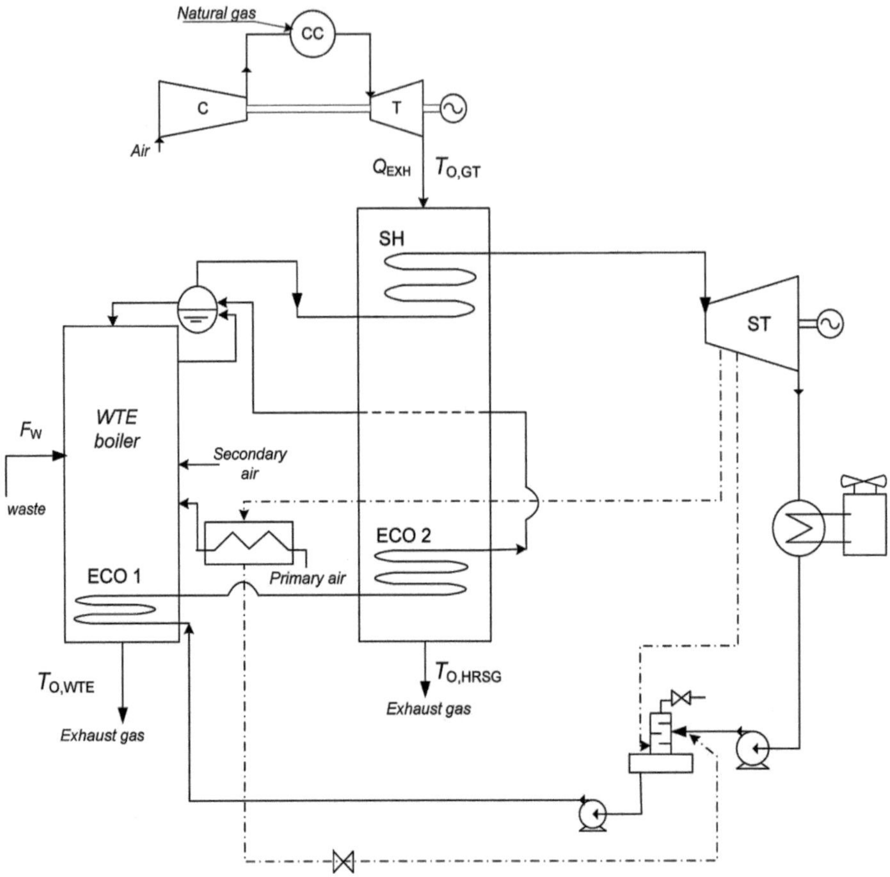

Fig. 6.14 Schematic of Layout_2 waste-to-energy *(WTE)*–GT integrated plant. *C* compressor, *CC* combined cycle, *T* turbine, *SH* superheater, *ST* steam turbine, *ECO* economizer

increase in water inlet temperature at ECO2, for a fixed value of Q_{EXH}, increases the amount of generated steam. However, once the optimum condition is reached, the two layouts reach the same maximum value, since the amount of generated steam is determined only by the thermal power available with waste, F_W.

However, corresponding to an optimum value, an increase in HRSG outlet temperature is noticed when compared to Layout_1. In this configuration, water enters ECO2 at a higher temperature. The *T–Q* diagram for the HRSG section of Layout_2 is shown in Fig. 6.16. About 71 % of the thermal power introduced in HRSG is used, with 60 % used in the SH section.

Figure 6.17 shows an alternative possibility to share water economization between the two subsystems, namely, Layout_2bis. In the considered configuration, water coming out of the DEA, is first fed to the ECO2 (placed in the HRSG), where part of the economization is performed which is completed in the ECO1 inside the WTE section.

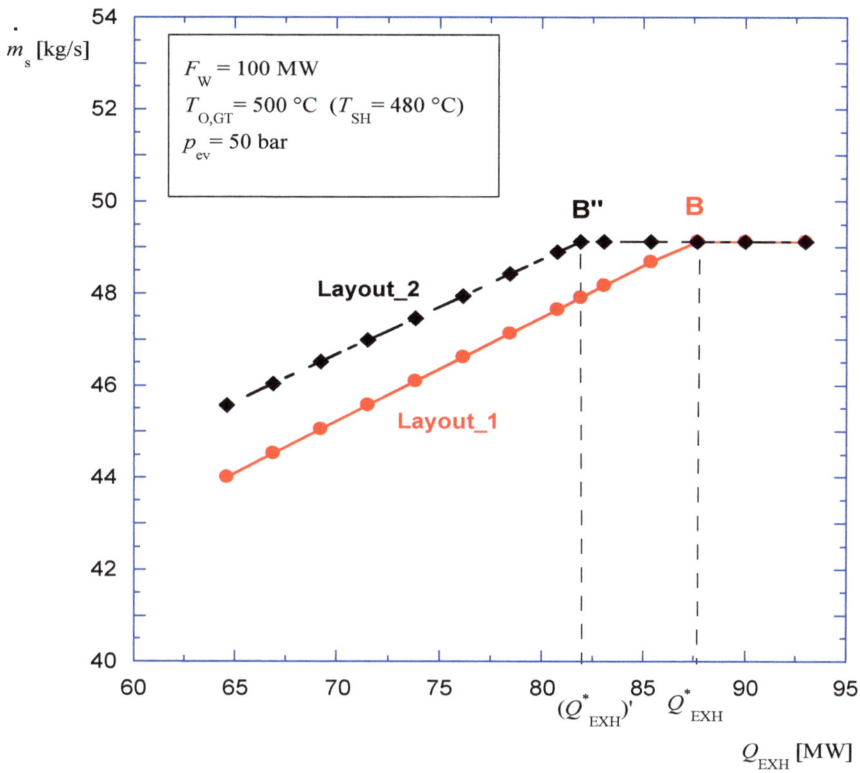

Fig. 6.15 Steam mass flow rate for Layout_1 and Layout_2 as a function of GT-discharged thermal power

The *T–Q* diagram relative to Layout_2bis is shown in Fig. 6.18. In this case, with respect to Layout_2, a lower HRSG-discharged temperature is obtained. Comparing the steam mass flow rate production across the proposed layouts (Fig. 6.19), it can be observed that the optimum condition of Layout_2bis (B′) is within point B″ (optimum value for Layout_2) and B (optimum value for Layout_1). Meanwhile, for all the proposed configurations, the maximum amount of generated steam mass flow rate remains the same.

The main results of Layout_2 and Layout_2bis are summarized in Table 6.6. It can be noticed that corresponding to the optimum conditions (point B′ in Fig. 6.19), Layout_2 reaches the minimum-allowed WTE exhaust gas temperature, while the HRSG outlet temperature turns out to be considerably higher than the minimum value allowed.

Layout_3 In the proposed layout, shown in Fig. 6.20, a parallel configuration between the ECO sections is evaluated. Water, coming out of the DEA, is divided into two streams: A fraction (*f*) goes into ECO2, the ECO section inside HRSG,

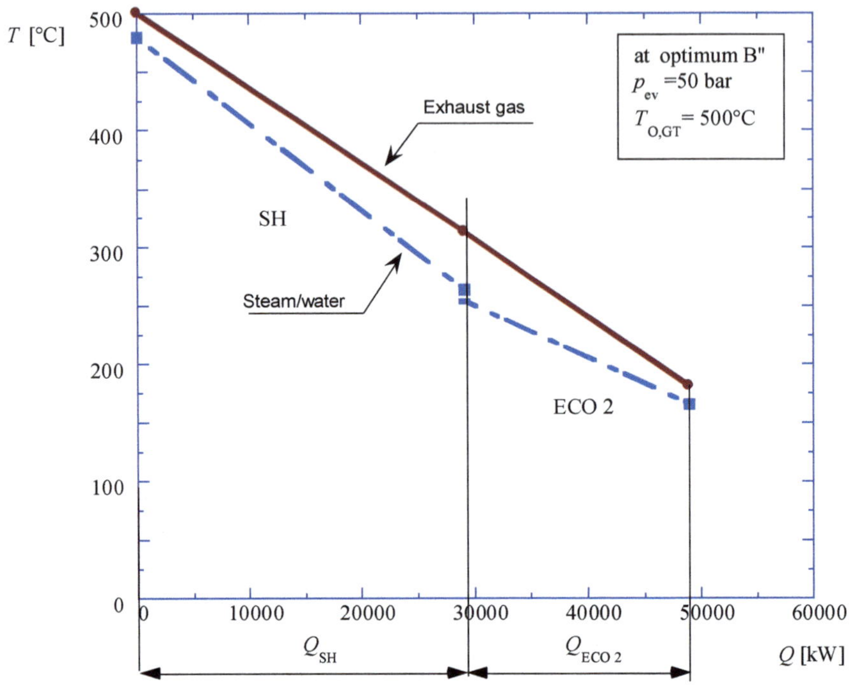

Fig. 6.16 Layout_2 *T–Q* diagram for the HRSG section. *SH* superheater, *ECO* economizer

while the remaining part is sent to ECO1, inside WTE. Before entering the WTE boiler, a mixer, combining both streams, is present.

Water mass flow rate splitting is adjusted to have similar mixer inlet temperatures (or ECO outlet temperatures) in trying to maximize the exploitation of the available thermal power, minimizing exhaust gas temperatures.

Corresponding to the optimum condition, for an evaporative pressure equal to 50 bar and a GT outlet temperature of 500 °C, the following water mass flow rate splitting has been found: 70 % of the total water mass flow is sent to ECO2 inside HRSG. Corresponding to the optimum Q^*_{EXH}, the assumed water mass flow rate splitting allows one to reach the same minimum water subcooling temperature difference of both streams. Below the optimum value (point B‴, Fig. 6.21), due to a decrease in Q_{EXH}, a slight decrease in *f* must occur, increasing the water mass flow rate economized in the WTE section.

As highlighted in Fig. 6.21, the proposed configuration allows a considerable decrease in Q^*_{EXH} (about 11 % decrease in Layout_1). Moreover, for $Q_{EXH} < Q^*_{EXH}$, Layout_3 achieves the best performance in terms of the generated steam mass flow rate.

Focusing on the *T–Q* diagram for Layout_3 (Fig. 6.22), a reduction in the distance between the exhaust gas and steam/water lines with respect to previous layouts can be observed; moreover, water mass flow rate splitting has been also optimized in order to obtain a parallelism between exhaust gas and water lines.

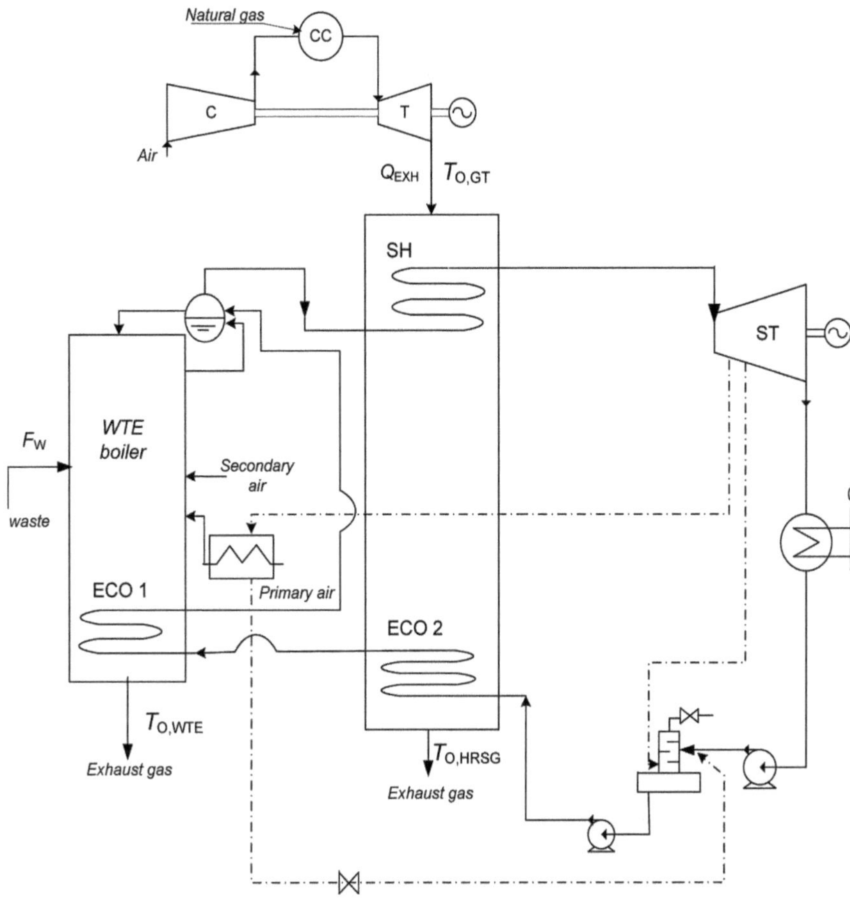

Fig. 6.17 Schematic of Layout_2bis waste-to-energy *(WTE)*–GT integrated plant. *C* compressor, *CC* combined cycle, *T* turbine, *SH* superheater, *ST* steam turbine, *ECO* economizer

The main results of Layout_3 are summarized in Table 6.7. It can be noticed that corresponding to optimum conditions, the WTE exhaust gas temperature is close to its minimum value ($T_{\mathrm{O,\,WTE}}$ equals about 168 °C).

Layout_4 Starting with Layout_2, the possibility of further exploiting the HRSG-discharged thermal power has been evaluated in the next proposed configuration, namely, Layout_4. Starting with the results of Layout_2, in terms of HRSG outlet temperature, the possibility to preheat waste combustion air has been investigated in this layout. At the exit of the ECO section, a heat exchanger has been introduced fed by HRSG-discharged thermal power. This solution allows eliminating the ST bleed necessary for air preheating.

Layout_4 refers to the preheating of only the primary combustion air up to the set temperature of 150 °C, while the temperature of the secondary combustion air is set at 50 °C. Preheating of both primary and secondary combustion air has been investi-

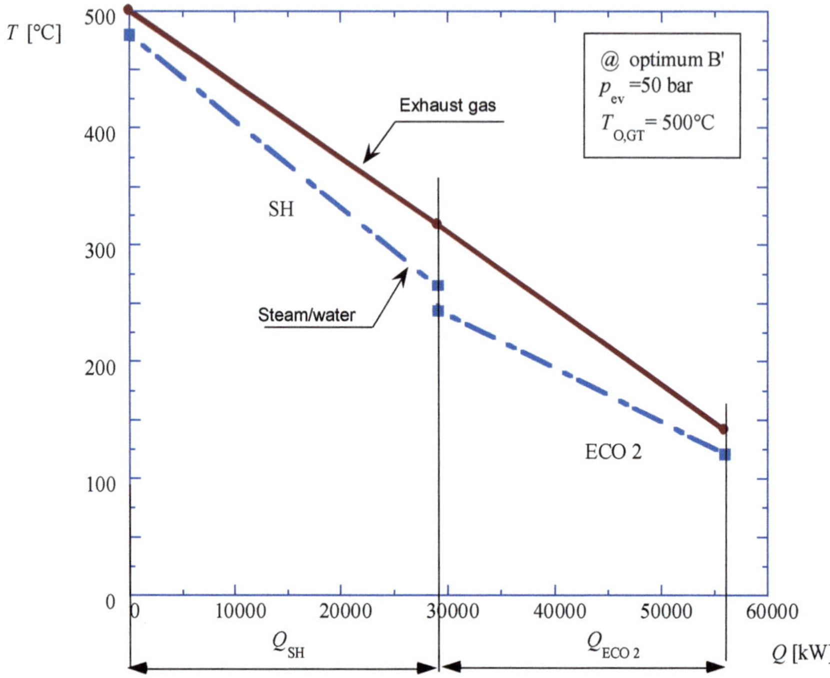

Fig. 6.18 Layout_2bis *T–Q* diagram for the HRSG section. *SH* superheater, *ECO* economizer

gated (namely, Layout_4_tot_air), where both temperatures (primary and secondary combustion air) have been set equal to 150 °C.

Focusing on Fig. 6.23, an increase in the maximum amount of steam mass flow rate has been observed in both cases, equal to about 1 and 3.5 %, respectively, for Layout_4 and Layout_4_tot_air. For both layouts, an increase in steam mass flow rate (points B^{IV}) is achieved with respect to Layout_1 and Layout_2. This can be explained considering the increase in combustion air temperature which, for the considered layouts, reaches 150 °C with an increase equal to 20 °C on previous configurations (see Table 6.2) and increases the reactants' enthalpy, therefore increasing the combustion temperature.

As shown in Table 6.8, the increase in maximum steam mass flow rate and the elimination of ST bleed for air preheating allow an increase in ST power output.

Layout_5 In the proposed layout (namely, Layout_5, shown in Fig. 6.24) the possibility of introducing an integral DEA inside the HRSG section has been analyzed, thus eliminating the need for a ST bleed to feed it. Water coming out of the condenser is sent to a pre-ECO placed in the HRSG, feeding the integral DEA (EVA DEA). Water mass flow rate out of the deaerator is then divided into two streams to be economized in parallel. Concerning water splitting between ECO1 and ECO2, similar considerations as those for Layout_3 were maintained.

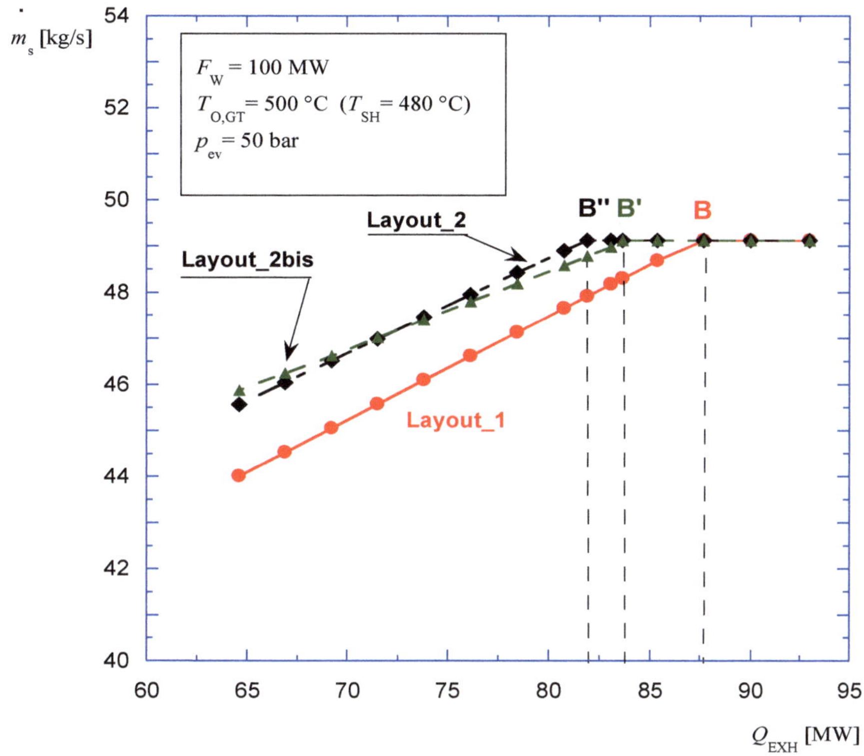

Fig. 6.19 Steam mass flow rate for Layout_1, Layout_2, and Layout_2bis as functions of GT-discharged thermal power

Table 6.6 Layout_2 and Layout_2bis results

	Layout_2	Layout_2bis
F_W (MW)	100	100
p_{ev} (bar)	50	50
$T_{O,GT}$ (°C)	500	500
T_{SH} (°C)	480	480
$\dot{m}_{EXH}$ (kg/s)	142	145
Q^*_{EXH} (MW)	81.91	83.64
Q^*_{EXH}/F_W (–)	0.82	0.84
$\dot{m}_{s,max}$ (kg/s)	49.12	49.12
$T_{O,HRSG}$ (°C)	181	141
$T_{O,WTE}$ (°C)	160	246
P_{ST} (MW)	45.76	45.76

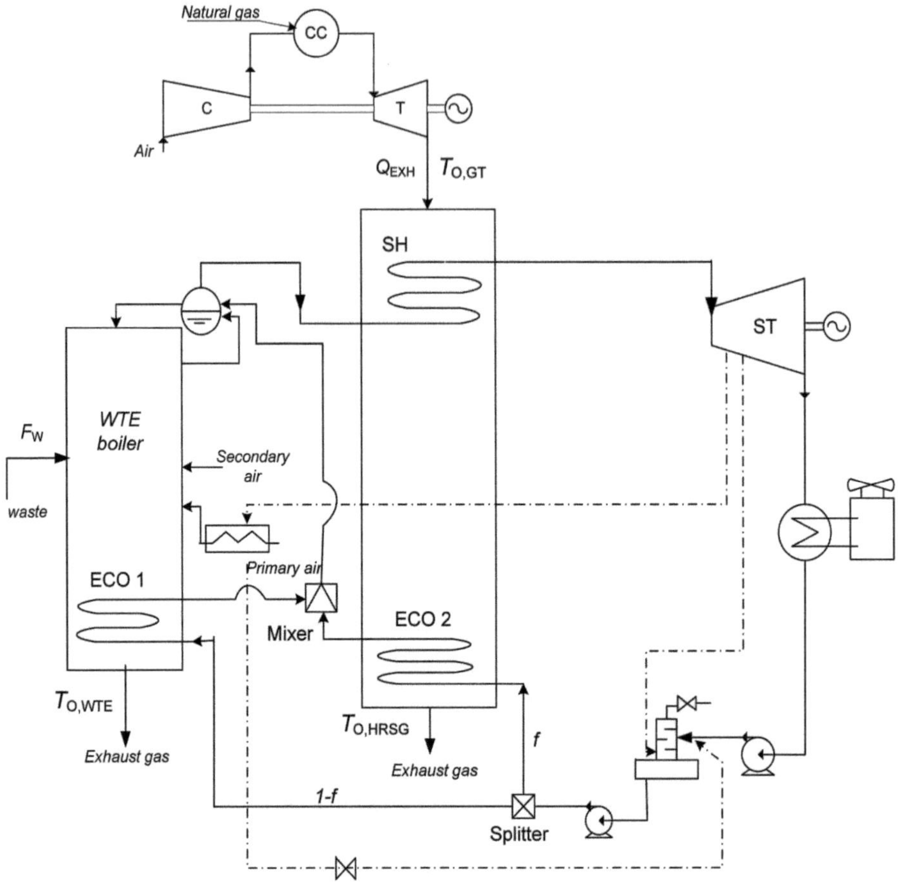

Fig. 6.20 Schematic of Layout_3 waste-to-energy *(WTE)*–GT integrated plant. *C* compressor, *CC* combined cycle, *T* turbine, *SH* superheater, *ST* steam turbine, *ECO* economizer

In Fig. 6.25, the steam mass flow rate produced in Layout_5 as a function of GT-discharged thermal power is shown. Comparing results of Layout_5 with other configurations, an increase in the optimum value of Q^*_{EXH} (point B^V) is found. Moreover, for $Q_{EXH} < Q^*_{EXH}$, Layout_5 has the lowest steam mass flow rate production. When the integrated system is operating below the optimum condition, water splitting between ECO2 and ECO1 deviates from the condition of point B^V (70 % of the total water mass flow to ECO2). A decrease in Q_{EXH} decreases the water mass flow rate that ECO2 can manage. Corresponding to the lowest considered value of Q_{EXH}, only 28 % of the total mass flow rate is fed to ECO2.

Focusing on the *T–Q* diagram in Fig. 6.26, the proposed configuration corresponding to the optimum plant match condition (point B^V) allows one to achieve the lowest HRSG outlet temperature ($T_{O,\,HRSG}$ is about 112 °C). Besides, the presence of the integral DEA inside HRSG increases the distance between the exhaust gases and

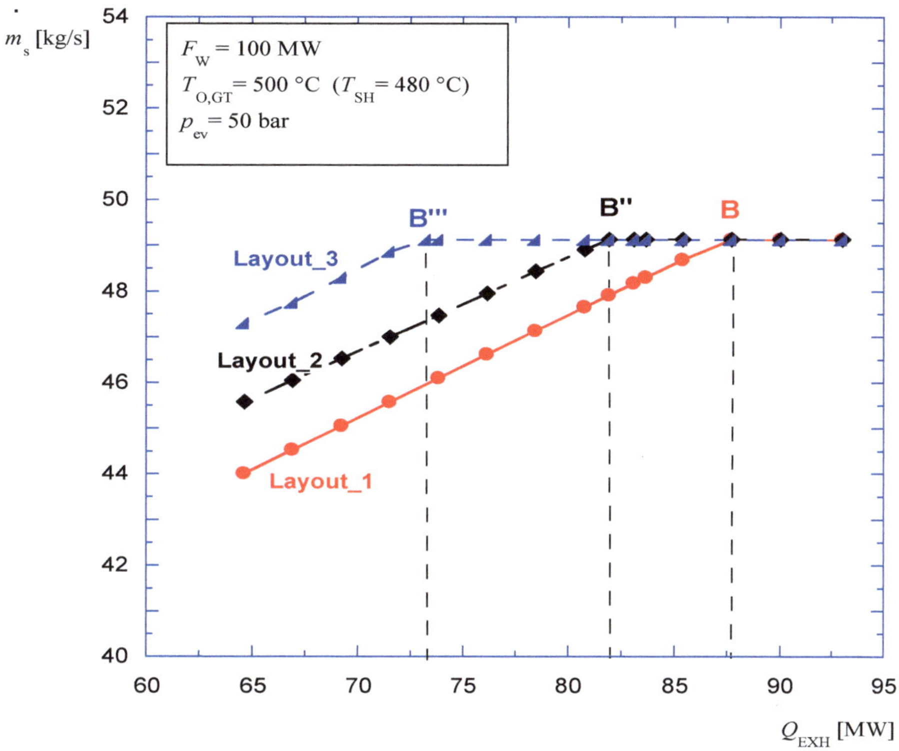

Fig. 6.21 Steam mass flow rate for Layout_1, Layout_2, and Layout_3 as functions of GT-discharged thermal power

steam/water lines. As shown in Table 6.9, the elimination of the ST bleed to feed the DEA allows an increase in ST power output.

Layout_6 In Layout_6, as shown in Fig. 6.27, there is a ST bleed to preheat feedwater (heat exchanger) before it enters the ECO. Out of the DEA, water economization is first performed in WTE (ECO1), then inside the HRSG section (ECO2).

The ST bleed for water preheating is set at 1.1 bar. This value is chosen so that the ST power output is not compromised and the *T–Q* diagram for HRSG is improved. Even in this layout, an integral DEA (EVA DEA) is present.

Starting with a DEA pressure of 2 bar, the working pressure has been changed to improve the HRSG *T–Q* profile (minimize the distance between exhaust gas and steam/water lines and decrease HRSG outlet temperature); an optimized pressure value of 3.5 bar has been found. The configuration obtained with the new value of the DEA pressure refers to Layout_6bis. *T–Q* diagrams relative to Layout_6 and Layout_6bis are shown in Figs. 6.28 and 6.29, respectively.

As shown in Fig. 6.30, the improved layout leads to a decrease in Q^*_{EXH} optimum value (point B^{VI} bis < B^{VI}) and to an increase in generated steam mass flow for $Q_{EXH} < Q^*_{EXH}$ (Table 6.10).

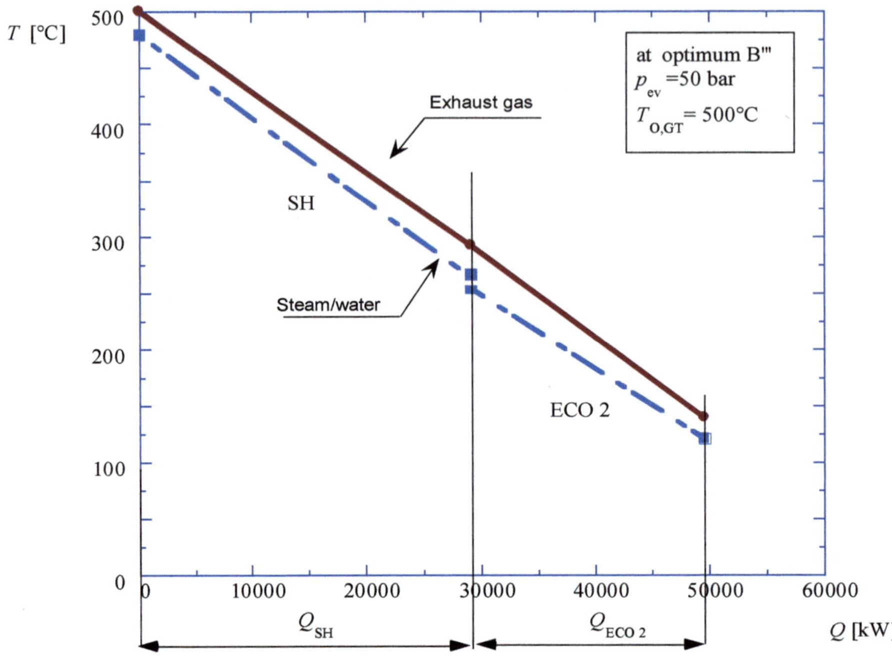

Fig. 6.22 Layout_3 *T–Q* diagram for HRSG section. *SH* superheater, *ECO* economizer

Table 6.7 Layout_3 results

Layout_3	
F_W (MW)	100
p_{ev} (bar)	50
$T_{O,GT}$ (°C)	500
T_{SH} (°C)	480
$\dot{m}_{EXH}$ (kg/s)	128
Q^*_{EXH} (MW)	73.83
Q^*_{EXH}/F_W (–)	0.74
$\dot{m}_{s,max}$ (kg/s)	49.12
$T_{O,HRSG}$ (°C)	140
$T_{O,WTE}$ (°C)	168
P_{ST} (MW)	45.76

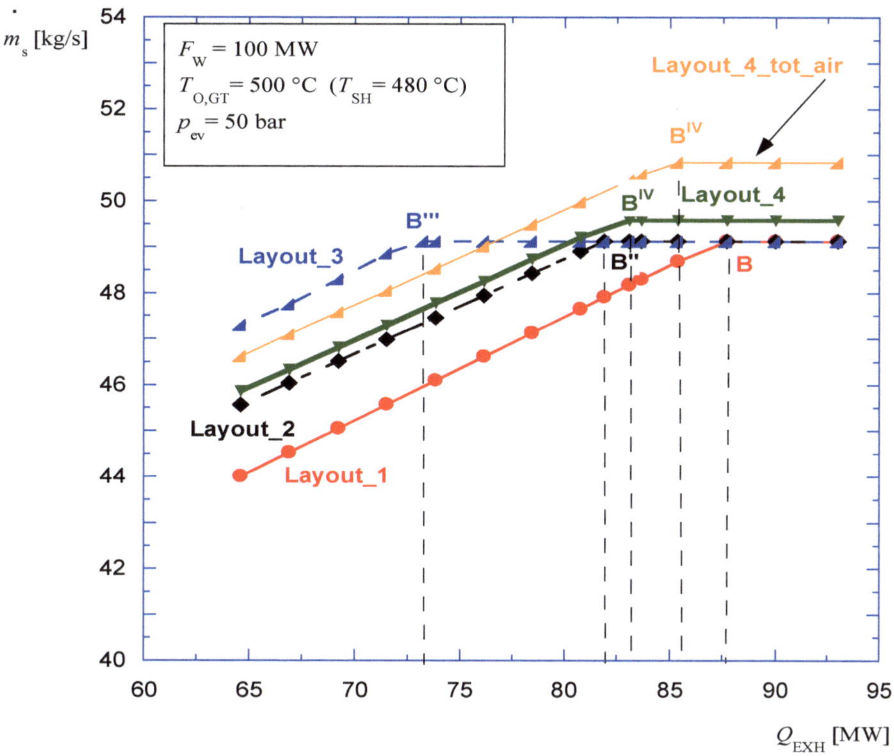

Fig. 6.23 Steam mass flow rate for different layouts as functions of GT-discharged thermal power

Table 6.8 Layout_4 and Layout_4_tot_air results

	Layout_4	Layout_4_tot_air
F_W (MW)	100	100
p_{ev} (bar)	50	50
$T_{O,GT}$ (°C)	500	500
T_{SH} (°C)	480	480
$\dot{m}_{EXH}$ (kg/s)	144	148
Q^*_{EXH} (MW)	83.06	85.37
Q^*_{EXH}/F_W (–)	0.83	0.85
$\dot{m}_{s,max}$ (kg/s)	49.57	50.84
$T_{O,HRSG}$ (°C)	158	144
$T_{O,WTE}$ (°C)	160	160
P_{ST} (MW)	46.69	47.88

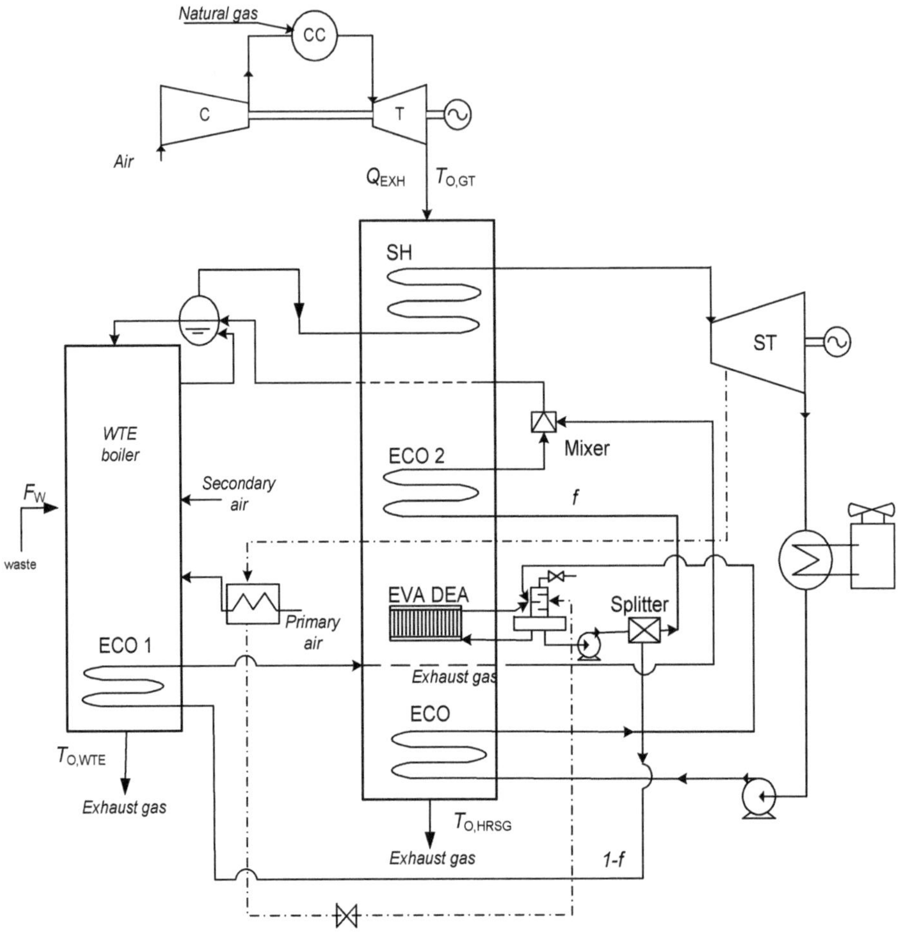

Fig. 6.24 Schematic of Layout_5 waste-to-energy *(WTE)*–GT integrated plant. *C* compressor, *CC* combined cycle, *T* turbine, *SH* superheater, *ST* steam turbine, *ECO* economizer, *EVA* evaporator, *DEA* deaerator

Layout_7 In the last proposed layout, Layout_7, Fig. 6.31, for a one-pressure-level integrated power plant, the possibility of generating a fraction of the total saturated steam in the HRSG section is investigated. Thus, a more complex configuration is proposed and investigated. The proposed layout can be interesting, in particular, considering the transient behavior of the system, LHV deviation from the design value, and the different system start-up times.

In Layout_7, an evaporative heat exchanger (EVA2) is placed in the heat recovery boiler unit; thus, a fraction of the total saturated steam is generated here. Water coming out of the condenser, before entering the deaerator, is divided into two streams: One part (*h*) is preheated in the ECO section inside HRSG, while the remaining (*1-h*) is sent to a heat exchanger fed by an ST bleed. Out of the DEA,

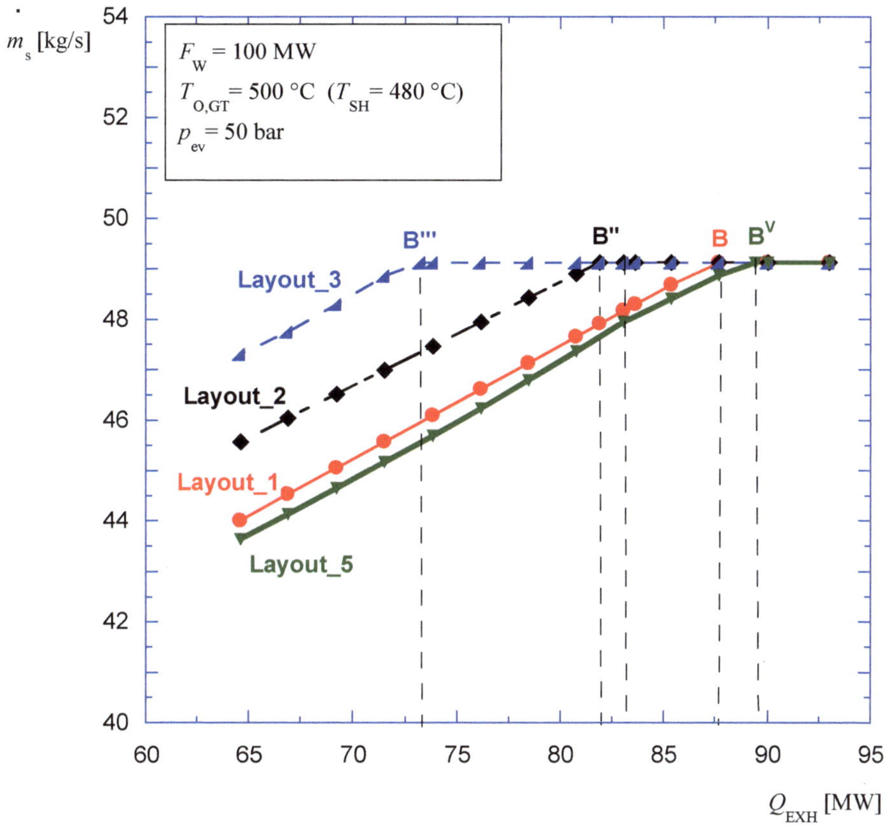

Fig. 6.25 Steam mass flow rate for Layout_1, Layout_2, Layout_3, and Layout_5 as functions of GT-discharged thermal power

water economization is performed in parallel: one part (*f*) is sent to the WTE ECO 1, while the other one (*1-f*) is sent to another ECO section inside HRSG (ECO 2); economized streams are mixed and then split again before they enter the EVA sections. Splitting of water entering the ECO sections is adjusted to exploit all the available heat inside the HRSG. Before entering the superheated heat exchanger, saturated steam streams are mixed together.

Contrary to the previously analyzed configurations, for Layout_7 there is no optimal condition in terms of Q_{EXH}; indeed, by increasing the value of Q_{EXH} the total steam mass flow rate, the sum of steam generated in the WTE boiler and in the HRSG section, always increases. However, a minimum value of GT-discharged thermal power is required to have steam generation inside the HRSG section. Figure 6.32 shows the steam mass flow rate generated in the WTE and HRSG sections plus the total amount as function Q_{EXH}; while in Fig. 6.33, the total amount of steam mass flow rate generated for Layout_7 is presented as a function

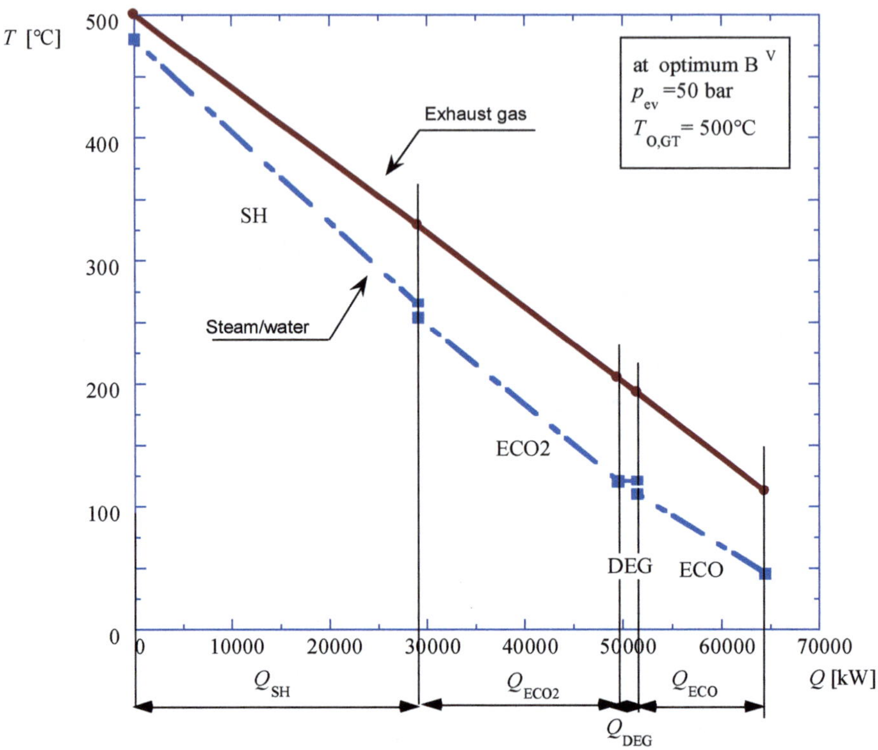

Fig. 6.26 *T–Q* diagram for the HRSG section for Layout_5. *SH* superheater, *ECO* economizer

Table 6.9 Layout_5 results

Layout_5	
F_W (MW)	100
p_{ev} (bar)	50
$T_{0,GT}$ (°C)	500
T_{SH} (°C)	480
$\dot{m}_{EXH}$ (kg/s)	155
Q^*_{EXH} (MW)	89.41
Q^*_{EXH}/F_W (–)	0.89
$\dot{m}_{s,max}$ (kg/s)	49.12
$T_{0,HRSG}$ (°C)	112
$T_{0,WTE}$ (°C)	168
P_{ST} (MW)	47.97

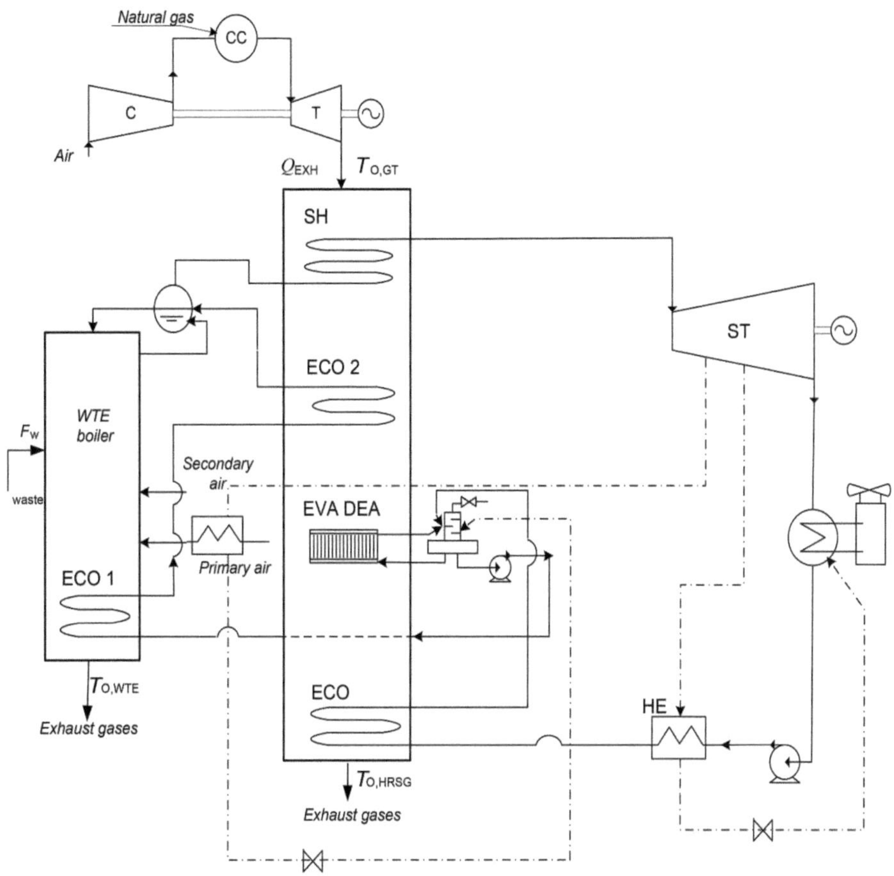

Fig. 6.27 Schematic of Layout_6 waste-to-energy *(WTE)*–GT integrated plant. *C* compressor, *CC* combined cycle, *T* turbine, *SH* superheater, *ST* steam turbine, *ECO* economizer, *EVA* evaporator, *DEA* deaerator, *HE* general heat exchanger

of GT-discharged thermal power. As highlighted in both figures, an increase in Q_{EXH} increases $\dot{m}_{s,HRSG}$; thus increasing the total amount ($\dot{m}_{s,TOT}$) of the generated steam.

The proposed configuration allows one to achieve the minimum HRSG outlet temperature (see the *T–Q* diagram of Fig. 6.34). Thus, with the above configuration, the GT-discharged thermal power is completely exploited. Moreover, as suggested in Table 6.11, the WTE outlet temperature reaches the minimum value allowed. The *T–Q* diagram shown in Fig. 6.34 and the numerical results reported in Table 6.11 refer to Q_{EXH} equal to 89.4 MW; since, for Layout_7, it is not possible to identify an optimum plant match condition, the highest value of GT-discharged thermal power previously found (see Layout_5 optimum condition) has been taken as a reference.

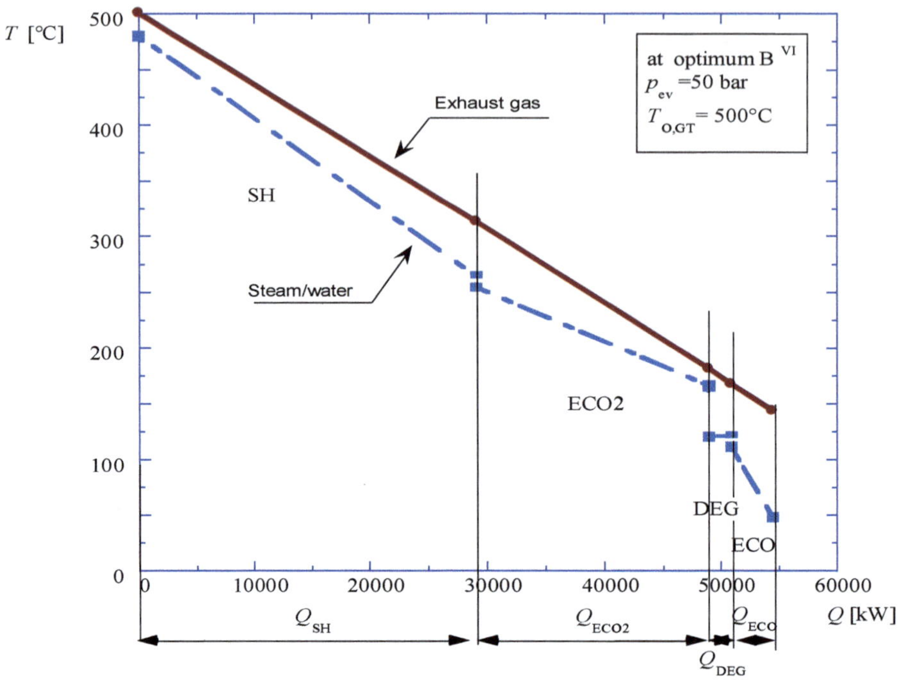

Fig. 6.28 *T–Q* diagram for the HRSG section in Layout_6. *SH* superheater, *ECO* economizer

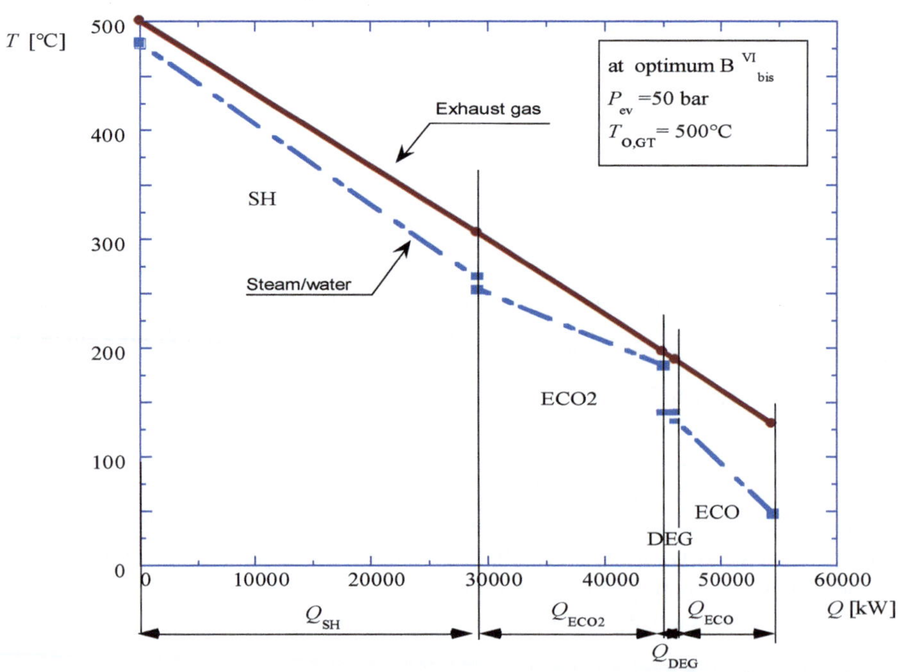

Fig. 6.29 *T–Q* diagram for the HRSG section in Layout_6bis. *SH* superheater, *ECO* economizer

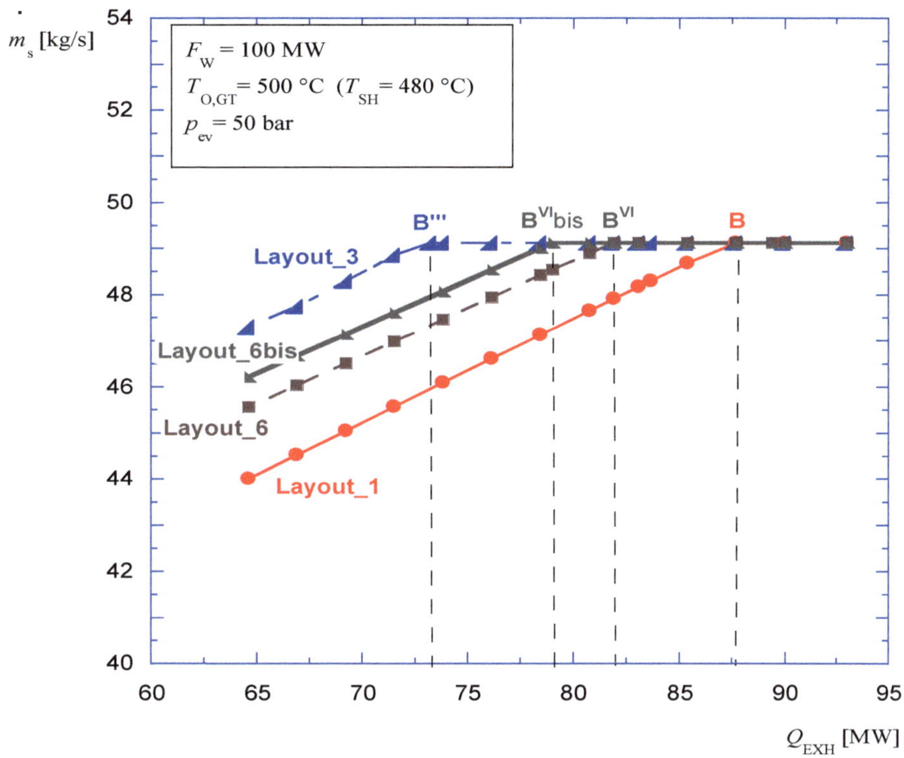

Fig. 6.30 Steam mass flow rate for Layout_1, Layout_3, Layout_6, and Layout_6bis as functions of GT-discharged thermal power

Table 6.10 Layout_6 and Layout_6bis results

	Layout_6	Layout_6bis
F_W (MW)	100	100
p_{ev} (bar)	50	50
$T_{O,GT}$ (°C)	500	500
T_{SH} (°C)	480	480
$\dot{m}_{EXH}$ (kg/s)	142	137
Q^*_{EXH} (MW)	81.91	79.02
Q^*_{EXH}/F_W (–)	0.82	0.79
$\dot{m}_{s,max}$ (kg/s)	49.12	49.12
$T_{O,HRSG}$ (°C)	143	130
$T_{O,WTE}$ (°C)	160	160
P_{ST} (MW)	46.72	46.72

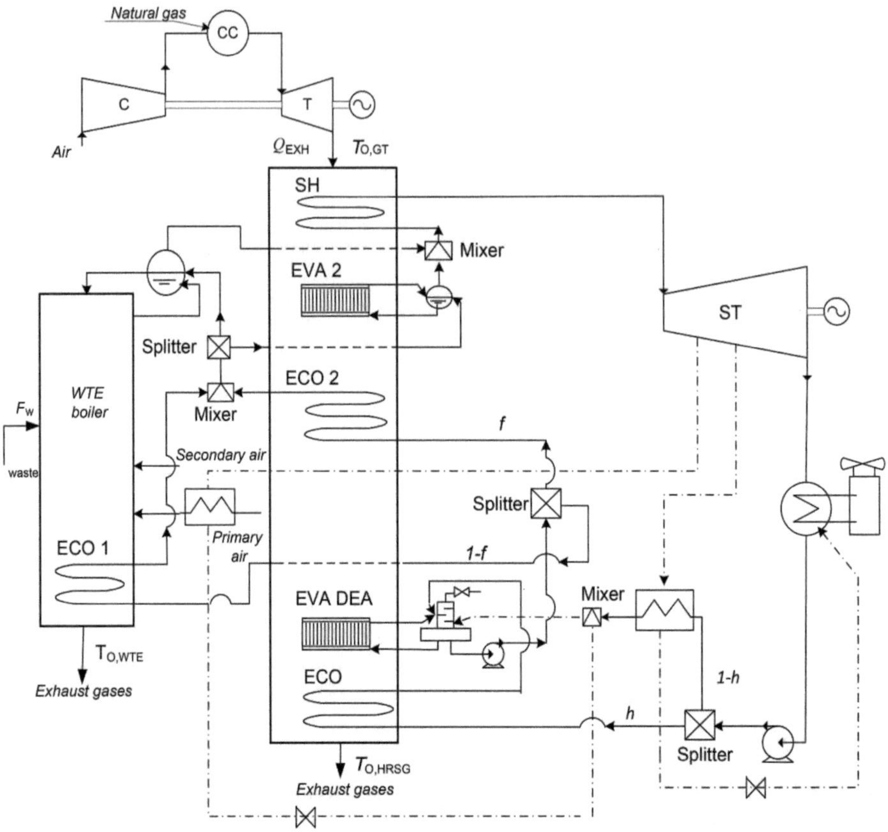

Fig. 6.31 Schematic of Layout_7 WTE–GT integrated plant. *C* compressor, *CC* combined cycle, *T* turbine, *SH* superheater, *ST* steam turbine, *ECO* economizer, *EVA* evaporator, *DEA* deaerator

6.5 Comparative Results of WTE–GT One-Pressure-Level Integrated Layouts

The results of all the proposed one-pressure-level WTE–GT integrated layouts are summarized in Figs. 6.35, 6.36, and 6.37, showing the maximum amount of steam mass flow rate, the ST power output, and the ratio of steam to waste mass flow rate, respectively, as a function of GT-discharged thermal power in correspondence to optimum plant match conditions (points B) identified in the previous section.

As previously discussed, for the proposed configurations, Layout_3 reaches the minimum value of Q^*_{EXH}. While Layouts 1, 2, 2bis, 5, 6, and 6bis attain the same maximum steam mass flow rate generation, Layouts 4, 4_tot_air, and 7 lead to an increase in the maximum amount of steam that a plant could produce.

As far as the ST power capacity is concerned, Layout_7 gives the best performance, followed by Layout_5 and Layout_4_tot_air.

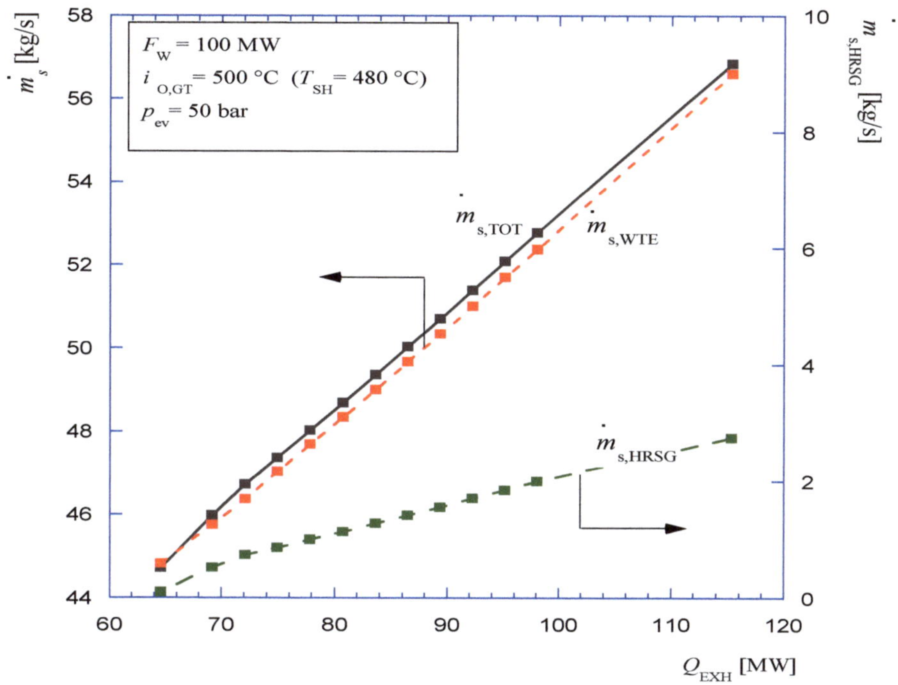

Fig. 6.32 Steam mass flow rate for Layout_7 as a function of GT-discharged thermal power

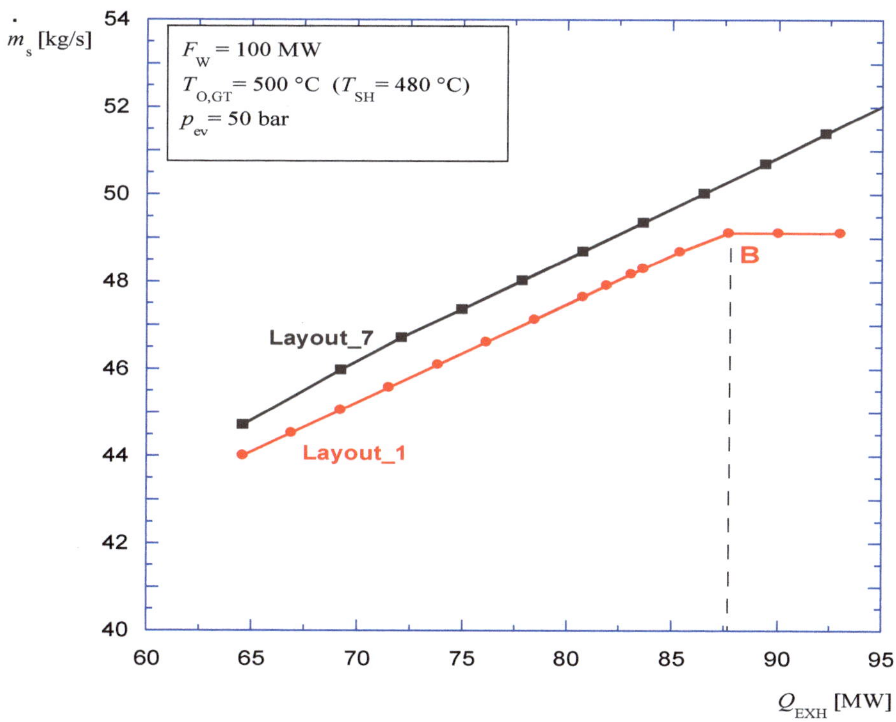

Fig. 6.33 Steam mass flow rate for Layout_1 and Layout_7 as a function of GT-discharged thermal power

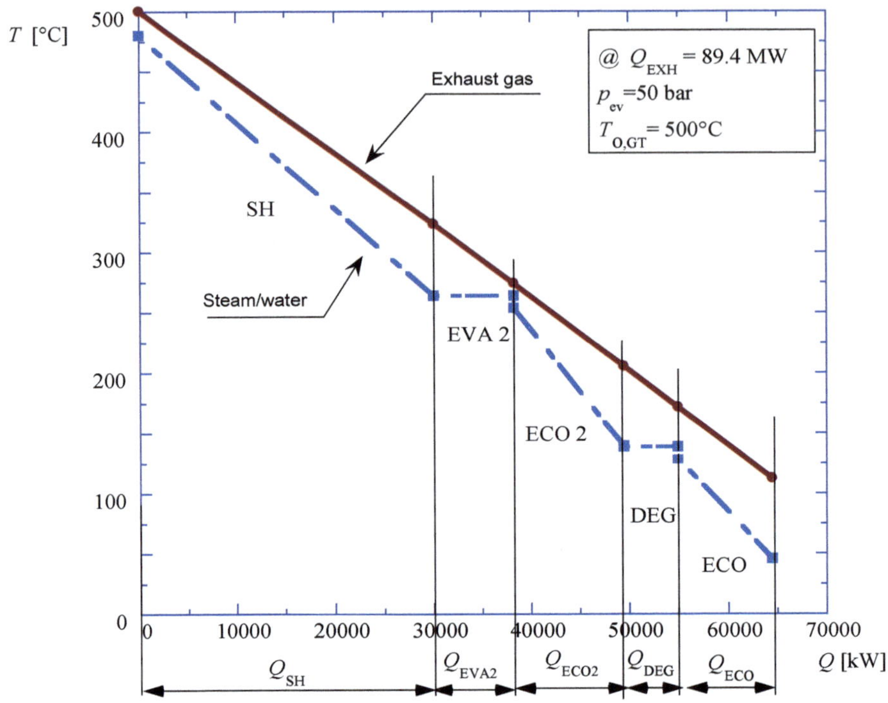

Fig. 6.34 *T–Q* diagram of HRSG section for Layout_7. *SH* superheater, *ECO* economizer, *EVA* evaporator

Table 6.11 Layout_7 results

Layout_7	
F_W (MW)	100
p_{ev} (bar)	50
$T_{O,GT}$ (°C)	500
T_{SH} (°C)	480
$\dot{m}_{EXH}$ (kg/s)	155
Q_{EXH} (MW)	89.41
Q_{EXH}/F_W (–)	0.89
$\dot{m}_{s,max}$ (kg/s)	50.70
$T_{O,HRSG}$ (°C)	112
$T_{O,WTE}$ (°C)	160
P_{ST} (MW)	48.95

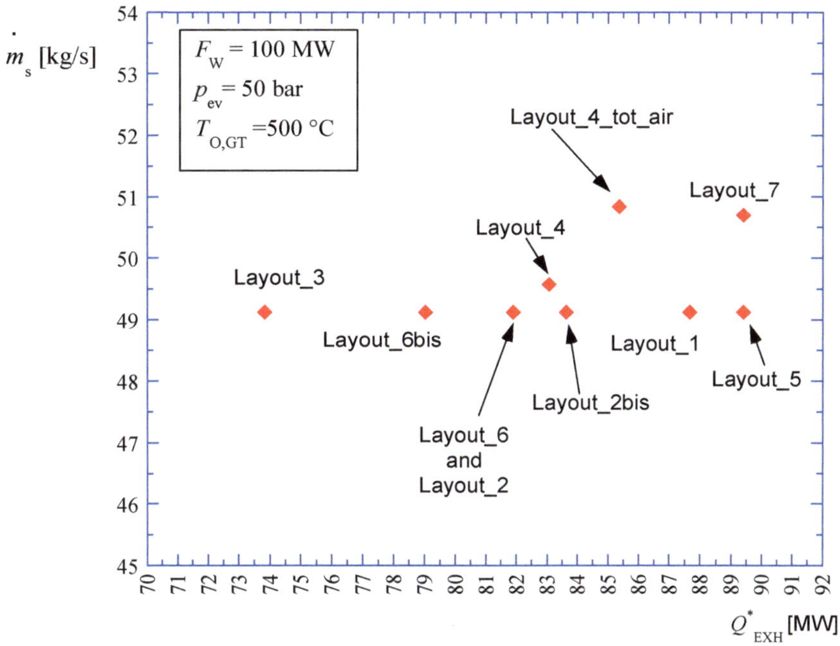

Fig. 6.35 Maximum steam mass flow rate (optimum condition, points B) as a function of GT-discharged thermal power

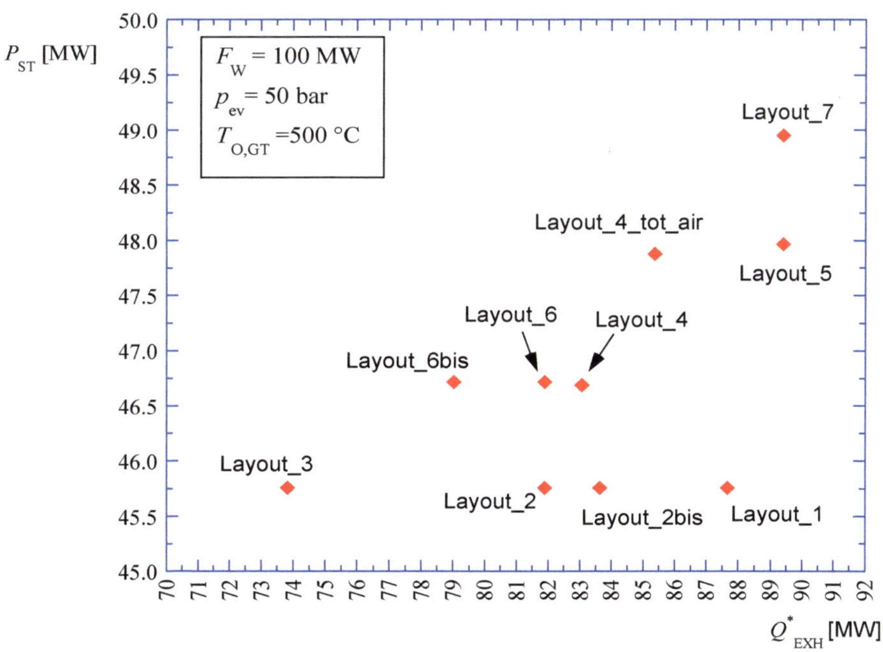

Fig. 6.36 Steam turbine power output as a function of GT-discharged thermal power

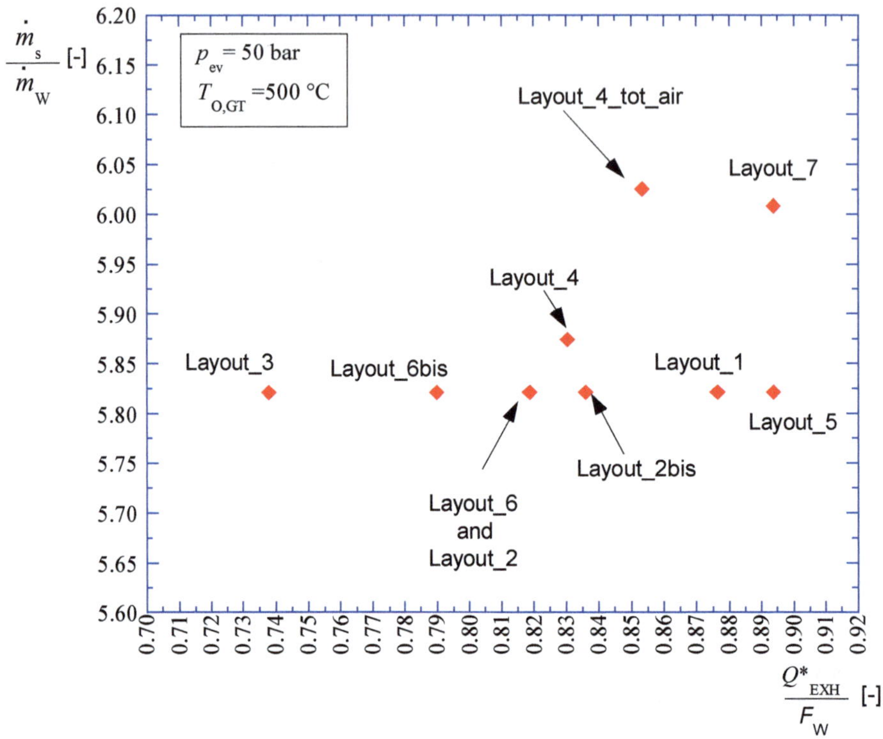

Fig. 6.37 Steam-to-waste mass flow rate ratio as a function of thermal power input ratio

The analysis shows that the steam mass flow rate produced by the integrated plant, for all the proposed layouts, is higher than that generated with a traditional WTE power plant fed with the same amount of waste. Figure 6.38 shows an increment in steam mass flow rate and ST power output for each investigated layout compared to a traditional WTE fed with the same thermal power input ($F_W = 100$ MW) and working at the same evaporative pressure and steam-superheated temperature equal to 50 bar and 480 °C, respectively. The steam mass flow rate production of a stand-alone WTE would be equal to about 32.09 kg/s, while the ST capacity would be equal to 29.7 MW; thus, increments range, for the considered cases, from 53 to 58 % for the steam mass flow and from 54 to 65 % for the ST capacity.

Numerical results in terms of integrated system efficiency and power output allocation to each input fuel will be discussed, in detail, in the next chapter.

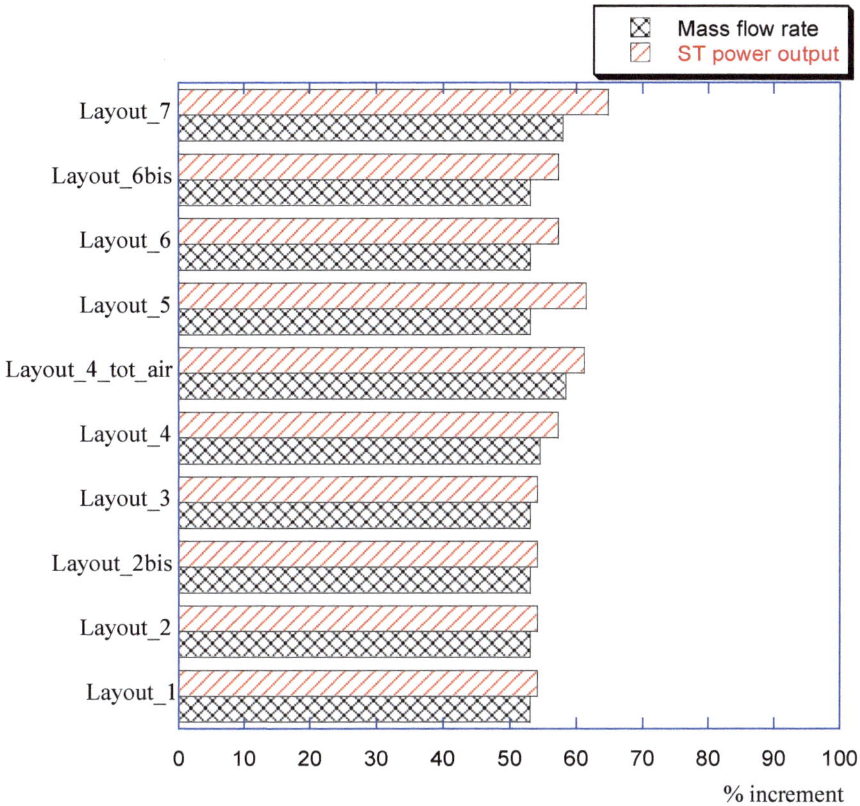

Fig. 6.38 Increments in steam mass flow rate with respect to a traditional stand-alone WTE power plant. *ST* steam turbine

References

1. Gate CycleTM by General Electric Energy. http://site.ge-energy.com/prod_serv/products/oc/en/opt_diagsw/gatecycle1.htm
2. Lozza G Turbine a gas e cicli combinati, Progetto Leonardo, 2° Edizione (In Italian)

Part IV
Performance and Efficiency Conversion Issues

Chapter 7
Performance Indexes and Output Allocation for Multi-fuel Energy Systems

In this chapter, the concept of conversion efficiency by exploring multi-fuel (MF) energy systems is discussed. An MF system can be defined as a system with various fuel energy inputs and useful product outputs. The difficulty in defining a performance index for MF systems lies in quantifying the contribution of each input fuel to the total output energy.

This chapter intends to make a few proposals and start a discussion that would help assess the MF system performance.

The conventional first law efficiency, normally used to assess the performance of a single-fuel (SF) system, can be applied, but it provides incomplete information about an MF system. The electric equivalent efficiency and the relative and overall MF synergy index (SI) concepts are introduced as the most significant performance indexes; these new performance indexes aim to evaluate the MF performance compared with reference SF scenarios. Proposed performance indexes are analyzed, discussed, and compared from a general point of view, identifying aspects of relation and the main differences. The introduced conversion efficiency indexes are then applied to a specific co-combustion power plant case. The influence of the reference scenarios, as it is highlighted in the chapter, is of great importance. The evaluation of the proposed indexes is of relevance for the formulation of a specific and unified theory about MF power systems, such as a waste-to-energy–gas turbine (WTE–GT) integrated power plant.

7.1 Context

A trend towards energy supply diversification, based on an energy infrastructure for which the feedstock supply no longer depends on a single fuel or network, is needed to prevent reliability reductions in the provided output energy that may occur in a market-oriented system. In addition, supply diversification provides more degrees of freedom in selecting the supply source while also taking into account the availability of storage from various sources [1]. The diversification of

L. Branchini, *Waste-to-Energy*, DOI 10.1007/978-3-319-13608-0_7

energy feedstock can also provide, as a consequence, the push towards MF energy systems, using different fuels in the same power plant. Another reason that makes the study particularly relevant is the increasing depletion of primary fossil fuel resources. Due to their depletion and the contextual increase of renewable sources [2, 3], MF energy systems are becoming attractive, and their penetration is increasing in the energy market [4, 5], especially within external combustion energy systems [6–9]. For example, MF energy systems based on co-firing of a fossil fuel together with a limited fraction of renewable or waste-derived fuel can be a strategy to use conventional power plants by introducing only modifications in the combustion section and/or limited challenges in the boiler redesign [10]. Moreover, MF conditions can also be realized in renewable-energy-based [11] or waste-to-energy (WTE) [12–14] power plants, which nevertheless need conventional fuels in more or less limited periods of their operation.

In SF energy systems, where input fuel energy is converted into useful output energy, the evaluation of conversion efficiency is a simple issue resulting from the comparison between the generated output and the introduced input. By-products of the conversion process, such as discharged heat, are disregarded and/or treated as waste. For MF energy systems, a standardized performance index that is able to evaluate both the overall conversion efficiency of the input into the total output and the relative influence on performance of the single input is still under research. The most problematic aspect concerning the MF system conversion efficiency is that different fuels with different characteristics and value all contribute to the generation of the overall system energy output. Therefore, it is hard to measure each individual input contribution to the global output. The problem of fuel valorization can become very useful from a practical point of view, for example when one of the input fuels of an MF power plant is renewable and the other is not. Therefore, it is important to understand which fraction of the output can be recognized as "clean," and subsequently opening it to incentives or special tariff regimes. Another problem can arise when the integration between two or more fuels (e.g., one renewable and another natural gas) is accomplished in a plant that, due to the MF arrangement, cannot achieve the combined-cycle-state-of-the-art efficiency values close to 60 %. In this case, the question that can be raised is: "Is the integrated MF arrangement energetically convenient?" Generally, from a theoretical point of view, it should be useful to identify comparison indexes, representing "best" or "reference" or at least theoretical limiting efficiency values in order to stimulate the technical development towards significant improvements in MF plants. This chapter puts forward some new definitions and discusses the conversion efficiency of an MF energy system from different points of view. Absolute and comparative performance indexes are presented to evaluate and compare conversion efficiency of an MF system with respect to SF systems fed with the same amount of input energy. Eventually, examples of MF systems are given to highlight their potential advantages in comparison with separate SF systems. Moreover, the analyzed cases show how the selection of the reference SF scenario can affect the convenience of the MF solutions.

7.2 Performance Evaluation of an MF Energy System

The complexity of the issues related to MF energy systems calls for synthetic indicators to characterize and define the performance of the system. In particular, the proposed approach considers the energy system as a black box. Using this black-box approach, it is possible to build an equivalent performance model on the basis of only input–output energy flows, without considering the detailed representation of the internal components.

For notation purposes, the various energy entries are calculated over a given time span (for instance, hourly, daily, annual) depending on the purpose of the study. Moreover, this discussion is restricted to energy systems in which flows are steady.

7.2.1 MF Energy System Arrangement

A general MF energy system is shown in Fig. 7.1 as a black box, fed by n fuel inputs (F_i is the ith associated energy), which produce a useful electric energy output *(E),* while rejecting a non-useful heat (Q_{nu}). Moreover, within the framework of a general-purpose approach, heat exchanges between the MF system and the external ambient or with other systems are taken into account in this study. Indeed, as schematically shown in Fig. 7.1, both input heat and output heat are considered. In particular, the system can receive additional input heat (Q_{in}), and it can deliver useful recoverable heat (Q_{rec}). The control surface, surrounding the MF energy system, can be applied to single components constituting the power plant or the whole system, to evaluate, respectively, the subsystem and/or the overall energy system conversion efficiency. The energy system input F_i can be multiple and from different types of sources such as (i) primary fuels, directly available in nature (e.g., natural gas, coal, oil, biomass, etc.) and (ii) non-primary fuels (energetically expensive and

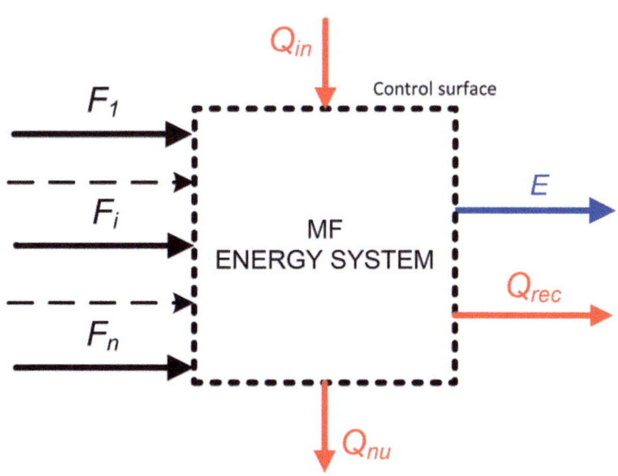

Fig. 7.1 Basic multi-fuel *(MF)* energy system

not directly available in nature, e.g., H_2, syngas, methanol, etc.). In more detail, in the case of primary fuels, the expression of F_i is the following:

$$F_i = m_i \cdot LHV_i, \tag{7.1}$$

where, for each input source, m_i represents the fuel mass flow and LHV_i represents the lower heating value per unit of mass flow at the ambient temperature (T_0) at which the ith reactant enters. For non-primary fuel input cases, the term F_i represents the primary fuel energy ideally necessary to produce the actual input. In the case of high-temperature input streams, the relating enthalpy content is given by the term Q_{in}.

7.2.2 Indexes for MF Energy System Performance Evaluation

In order to evaluate the efficiency of an MF energy system, different performance indicators can be used, as described below.

7.2.2.1 First Law Efficiency

As a result of the laws of thermodynamics, by comparing useful output energy and input energy, the following index can be introduced:

First law efficiency:

$$\eta_I = \frac{E + Q_{rec}}{\sum_{i=1}^{n} F_i + Q_{in}}. \tag{7.2}$$

According to this definition, the useful output heat has the same importance as the electric output. This definition can be misleading as it does not consider the actual value of heat at different temperature levels.

7.2.2.2 Electric Equivalent Efficiency

In order to consider the quality of the useful input and output heat, including their contribution to the products, the electric equivalent of heat ε must be considered. Thus, in the proposed electric equivalent efficiency definition, the useful heat contribution is expressed as $+(\varepsilon_{rec} \cdot Q_{rec})$, while, with the same methodology, the inlet heat contribution can be expressed as $-(\varepsilon_{in} \cdot Q_{in})$ and can be considered a negative electric equivalent term; ε_{rec} and ε_{in} are the electric equivalent conversion factors of the heat terms, which depend on the temperature of the heat-exchanging

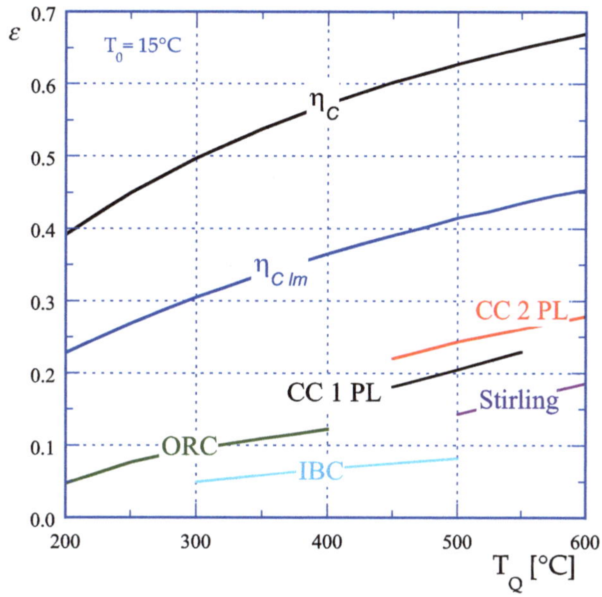

Fig. 7.2 Electric equivalent of heat, ε, as a function of the temperature level, T_Q, of the heat source for different heat recovery systems. *CC2PL* combined cycles with two-pressure-level arrangement, *CC1PL* combined cycles with one-pressure-level arrangement, *ORC* organic Rankine cycles, *IBC* inverted Brayton cycles

process. Therefore, the MF performance can be evaluated according to the following definition:

Electric equivalent efficiency:

$$\eta_{EE} = \frac{E + \varepsilon_{rec} \cdot Q_{rec} - \varepsilon_{in} \cdot Q_{in}}{\sum_{i=1}^{n} F_i} = \frac{U}{\sum_{i=1}^{n} F_i} \tag{7.3}$$

where U represents, with a compact notation, the total useful electric output, thus including the equivalent heat contributions. Different possibilities can be chosen to quantify the value of ε for each heat contribution, considering that the upper theoretical limit to ε is represented by the Carnot efficiency (η_C). Figure 7.2 shows the trend of ε versus temperature level T_Q of the heat source for different existing and innovative energy conversion systems, namely, combined cycles with two-pressure-level arrangement (CC 2 PL), combined cycles with one-pressure-level arrangement (CC 1 P L), organic Rankine cycles (ORC) with octamethyltrisiloxane (MDM) as the working fluid [7], Stirling engines, and inverted Brayton cycles (IBC) [6], where η_{Clm} is the efficiency of an ideal reversible recuperation cycle with limited capacity of the heat source.[1]

[1] An isentropic expansion and an isothermal compression is made up of an isobaric heat absorption during the process of heat exchange. The hot source temperature decreases from T_Q to T_0.

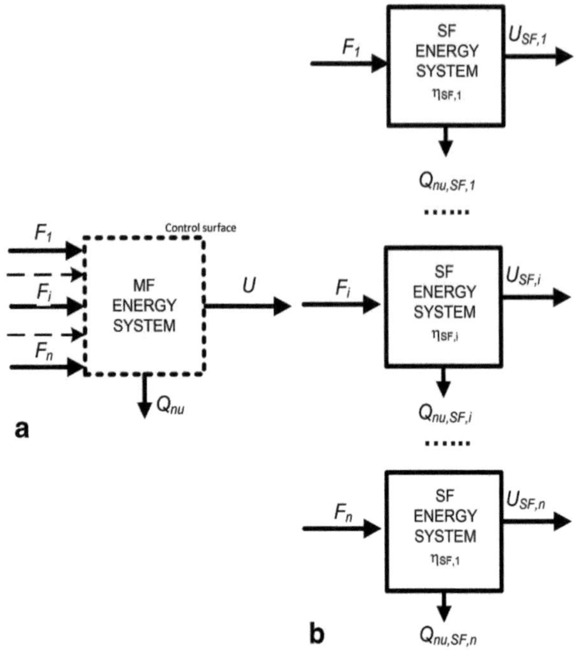

Fig. 7.3 **a** Multi-fuel *(MF)* energy system versus **b** single-fuel *(SF)* energy system

7.2.2.3 Relative SI

A different approach to measure the MF system performance is to quantify improvements obtained with the MF integrated system in comparison with separate exploitation of the various fuel inputs. Thus, a comparative energy conversion scenario must be taken into account. This approach is clearly described in Fig. 7.3, where the MF energy system (Fig. 7.3a) and the comparative scenario (Fig. 7.3b) are presented.

The improvement in the integrated MF configuration, in comparison with the separate and SF ones, is evaluated as being equal to all of the *i*th fuel energy input. A comparative index, named relative synergy of the *i*th source SI_i, is introduced here. This index tries to express the benefit of the *i*th fuel in the MF integrated plant by excluding the effects of all the other fuels and using the reference scenario to measure the useful output. In order to quantify the contribution of the *i*th source, the idea is to subtract from the total MF output (U) the output of all the reference SF systems ($U_{SF,j}$ in Fig. 7.3) except the *i*th is $\sum_{j=1, j\neq i}^{n} U_{SF,j}$. This quantity allows one to calculate SI_i.

Relative SI:

$$SI_i = \frac{U - \sum_{j=1, j\neq i}^{n} U_{SF,j}}{F_i} \tag{7.4}$$

To evaluate the output produced by an SF system $U_{SF,j}$, it is necessary to identify the specific SF reference system and in particular the characteristics of conversion efficiency. Thus, the generic $U_{SF,j}$ can be expressed as

$$U_{SF,j} = \eta_{SF,j} \cdot F_j, \tag{7.5}$$

where $\eta_{SF,j}$ represents the generic jth SF system conversion efficiency. Therefore, Eq. (7.4) can be rewritten as

$$SI_i = \frac{U - \sum_{j=1, j \neq i}^{n} (\eta_{SF,j} \cdot F_j)}{F_i}. \tag{7.6}$$

The selection of an appropriate value of each $\eta_{SF,j}$ can severely affect the advantages or disadvantages of the integrated system, as shown by a numerical example. Several possibilities of estimating the reference efficiency can be considered. For example, the "Best Available Technologies (BAT) Reference Document" (BREF) [15] can be used. For some of the most common fuels, the reference efficiencies are reported in [15] (Table 7.1) as a function of the combustion technology or the plant type.

7.2.2.4 MF SI

To evaluate the overall MF system performance and to give information about the synergy effect of all the input fuels in comparison with the SF systems of Fig. 7.3, the following index is also introduced:

MF SI:

$$SI = \frac{U - \sum_{i=1}^{n} U_{SF,i}}{\sum_{i=1}^{n} U_{SF,i}}. \tag{7.7}$$

Evidently, when $SI > 0$, the integrated MF energy system is better than the sum of the SF ones; the synergy effect due to the configuration of the plant and the contextual fuel exploitation increases the useful output in comparison with the overall output of the SF reference scenario. Otherwise, when $SI \leq 0$, the integrated configuration is equal to or worse than the collection of separate and SF systems.

The SI evaluates the overall benefit of the exploitation of all the input streams inside the integrated system differently from the relative SI, even though it is difficult to highlight and quantify the contribution of each SF source in terms of useful energy output. Nevertheless, it is possible to correlate both SI and η_{EE} to the ensemble of SI_i, according to the following equations:

Table 7.1 Efficiency associated with BAT for different fuels as a function of the combustion technique [15]

Fuel	Combustion technique	η_{SF} (%)	
		New plants	Existing plants
Coal	Pulverized combustion	43–47	35–40
	Fluidized bed	>41	
	Pressurized fluidized bed combustion	>42	
Biomass	Grate firing	20	–
	Spreader-stoker	>23	
	Fluidized bed combustion	>28–30	
Lignite	Pulverized combustion	42–45	35–40
	Fluidized bed	>40	
	Pressurized fluidized bed combustion	>42	
Peat	Fluidized bed combustion	>28–30	–
Plant type		*η_{SF} (%)*	
		New plants	*Existing plants*
Gas turbine		36–40	32–35
Gas engine		38–45	–
Gas-fired boiler		40–42	38–40
CC with or without supplementary firing (HRSG) for electricity generation		54–58	50–54

CC combined cycle, *HRSG* heat recovery steam generator

$$\mathrm{SI} = \frac{1}{n}\left(\frac{\sum_{i=1}^{n}\mathrm{SI}_i \cdot F_i}{\sum_{i=1}^{n}\eta_{\mathrm{SF},i} \cdot F_i} - 1\right) \tag{7.8}$$

$$\eta_{EE} = \frac{1}{n}\left(\sum_{i=1}^{n} SI_i \cdot x_i + (n-1)\sum_{i=1}^{n}\eta_{SF,i} \cdot x_i\right), \tag{7.9}$$

where $x_i = F_i / \sum_{i=1}^{n} F_i$ represents a fuel input weighting factor of the *i*th relative SI.

Due to the SI_i definition, when $\mathrm{SI} > 0$ (i.e., $U > \sum_{i=1}^{n} U_{\mathrm{SF},i}$) the following inequality occurs:

$$\sum_{i=1}^{n}\mathrm{SI}_i\, F_i > U > \sum_{i=1}^{n} U_{\mathrm{SF},i}, \tag{7.10}$$

which means that, in this case, the sum of the energies associated with the various SI_i is higher than the sum of the output energies of the SF reference scenario.

7.2.3 *Useful Output Allocation to Each ith Fuel*

Besides the issue of conversion efficiency assessment of an MF energy system, there is another aspect with practical implications to be solved: What is the contribution (U_i) of each fuel input to the total output (U)? How can it be properly quantified?

Different approaches can be used to allocate output energy to different fuels.

7.2.3.1 Allocation Approach #1

The first proposed approach for output allocation to fuel input consists in using the following equation:

$$U_i = U \cdot \frac{F_i}{\sum_i^n F_i} = \eta_{EE} \cdot F_i \,. \tag{7.11}$$

Thus, according to this approach, each U_i is simply proportional to the *i*th fuel input. As shown in the numerical example, this approach can be misleading when input fuels have different characteristics and value because, using Eq. (7.11), all fuels have the same weighting factor (namely, η_{EE}).

7.2.3.2 Allocation Approach #2

An alternative evaluation method is proposed here to account for the actual conversion efficiency gain of each fuel in comparison with the SF reference scenario; the new allocation formula is

$$U_i = U \cdot \frac{SI_i \cdot F_i}{\sum_i^n SI_i \cdot F_i} \,. \tag{7.12}$$

According to this definition, each fuel input is weighted via the corresponding SI_i term. In this way, the input fuel with higher conversion efficiency in comparison with the SF reference scenario (i.e., higher SI_i) can be identified with making a higher contribution to the global output. Thus, this method privileges the fuel which is better exploited in the MF system in comparison with the SF scenario, whereas Approach #1 does not include such reference to the SF scenario.

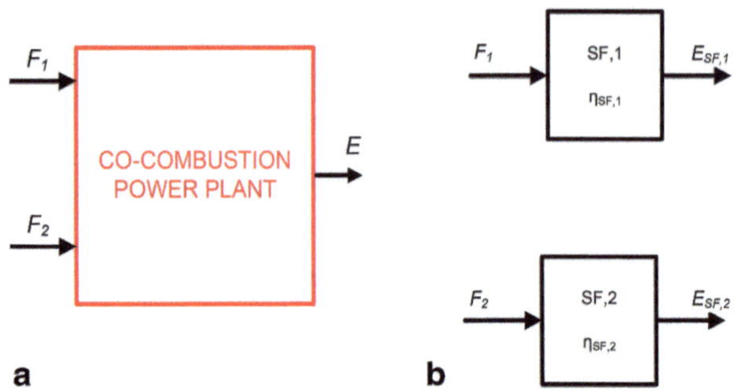

Fig. 7.4 Two-fuel co-combustion power plant (**a**) versus separate combustion energy systems (**b**)

The various efficiency definitions and allocation approaches can be applied to obtain useful information about the system convenience under different operating conditions or to compare different plant design arrangements. An application is presented in the following section.

7.3 Application Example: Two-fuel Co-combustion Power Plant

A specific case in which the above general methodology for MF systems performance evaluation can be applied is the simple case of co-combustion of two different fuels in a generic power plant. The scheme of the system is shown in Fig. 7.4a, where F_1 and F_2 are the fuel input energy values. In this case, the reference separate energy systems, exploiting the same two input fuels, are considered, as shown in Fig. 7.4b.

In particular, the co-combustion power plant produces the useful electric output E, while the two separate reference SF systems produce, respectively, $E_{\mathrm{SF},1}$ and $E_{\mathrm{SF},2}$. In this co-combustion case, the first law efficiency (also coincident with η_{EE}) according to Eq. (7.2) is

$$\eta_I = \frac{E}{F_1 + F_2}. \tag{7.13}$$

Evaluated in comparison with the reference scenario, the MF SI according to Eq. (7.7) is

$$\mathrm{SI} = \frac{\mathrm{E} - (E_{\mathrm{SF},1} + E_{\mathrm{SF},2})}{(E_{\mathrm{SF},1} + E_{\mathrm{SF},2})}. \tag{7.14}$$

Finally, the relative SI referred to the two separate fuel inputs according to Eq. (7.6) are

$$\mathrm{SI}_1 = \frac{E - E_{\mathrm{SF},2}}{F_1} = \frac{E - F_2 \cdot \eta_{\mathrm{SF},2}}{F_1} \tag{7.15}$$

$$\mathrm{SI}_2 = \frac{E - E_{\mathrm{SF},1}}{F_2} = \frac{E - F_1 \cdot \eta_{\mathrm{SF},1}}{F_2}, \tag{7.16}$$

where $\eta_{\mathrm{SF},1}$ and $\eta_{\mathrm{SF},2}$ are the efficiencies of the two reference SF separate plants. In order to obtain a global energy benefit from the co-combustion, the condition $\mathrm{SI} > 0$ leads to

$$E > (E_{\mathrm{SF},1} + E_{\mathrm{SF},2}). \tag{7.17}$$

This condition can be rearranged as a function of the co-combustion first law efficiency and of the relative SI of the two reference SF systems:

$$F_1(\eta_I - \eta_{\mathrm{SF},1}) + F_2(\eta_I - \eta_{\mathrm{SF},2}) > 0, \tag{7.18}$$

which is, first of all, verified when $\eta_I > \eta_{\mathrm{SF},1}$ and $\eta_I > \eta_{\mathrm{SF},2}$; otherwise, for $\eta_{\mathrm{SF},2} > \eta_{\mathrm{SF},1}$, the condition (7.18) can be also verified if the following relation is satisfied:

$$\frac{F_2}{F_1} < \frac{\eta_I - \eta_{SF,1}}{\eta_{SF,2} - \eta_I}. \tag{7.19}$$

This equation leads to a simple link between the co-combustion fuel ratio F_2 / F_1, the SF reference efficiencies, and a minimum first law conversion efficiency value η_{I_min}, which is required to have energy convenience (i.e., $\mathrm{SI} \geq 0$). This link is shown for two different reference SF efficiency cases (Fig. 7.5) corresponding to typical biomass and coal reference plants, with high- and low combustion technology levels, respectively, according to the values reported in Table 7.1, where fuel F_1 represents biomass and F_2 represents coal. The figure clearly shows that the convenience of the co-combustion plant is linked to the choice of the reference scenarios. The graph shows that to have convenience (i.e., a synergic effect) the first law efficiency of the co-combustion MF system must be higher than the weighted average of the two SF systems' efficiency, using the input fuel energy terms as weights.

Finally, in this co-combustion case the global convenience condition, that is $\mathrm{SI} > 0$ can be linked to the relative SI of each single fuel as follows:

$$\mathrm{SI} > 0 \Rightarrow (E - E_{\mathrm{SF},1} > E_{\mathrm{SF},2}) / F_2 \Rightarrow \mathrm{SI}_2 > \eta_{\mathrm{SF},2} \tag{7.20}$$

$$\mathrm{SI} > 0 \Rightarrow (E - E_{\mathrm{SF},2} > E_{\mathrm{SF},1}) / F_1 \Rightarrow \mathrm{SI}_1 > \eta_{\mathrm{SF},1}. \tag{7.21}$$

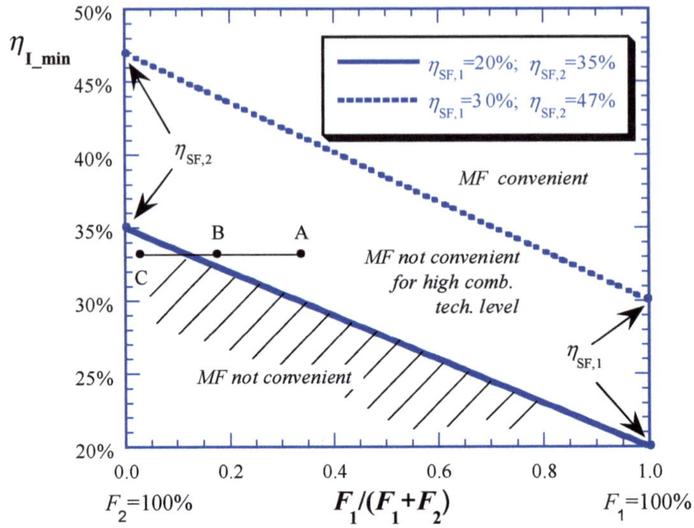

Fig. 7.5 Minimum total efficiency to have SI = 0 for a biomass–coal typical co-combustion power plant *versus* $F_1/(F_1+F_2)$. *MF* multi-fuel

These two conditions basically affirm that in the co-combustion plant both fuels are exploited better than in the two reference SF plants. Thus, the condition of relative-SI for each fuel entering the MF system higher than the corresponding SF system efficiency is equivalent to global MF convenience in terms of total output energy.

Table 7.2 shows numerical data of three co-combustion cases (corresponding points A, B, and C shown in Fig. 7.5) of biomass with coal used in the same power plant, with given size and first law efficiency; two MF cases (A and B) show synergy convenience and one (C) is not convenient. Due to inconvenience of case C ($SI < 0$), the evaluation of relative SIs and output allocation Approach #2 does not give meaningful results in this case.

The convenience of co-combustion is strictly related to the ratio between the two fuels and the different reference conversion efficiencies. The contribution of each input fuel to the total output is calculated according to the two above-mentioned approaches. In particular, using Approaches #1 and #2, the following expressions are obtained:

Approach #1:

$$E_1 = E \cdot \frac{F_1}{F_1 + F_2}; E_2 = E \cdot \frac{F_2}{F_1 + F_2} \tag{7.22}$$

Approach #2:

$$E_1 = E \cdot \frac{F_1 \cdot SI_1}{F_1 \cdot SI_1 + F_2 \cdot SI_2}; E_2 = E \cdot \frac{F_2 \cdot SI_2}{F_1 \cdot SI_1 + F_2 \cdot SI_2} \tag{7.23}$$

Table 7.2 Example of MF co-combustion performance

MF size and efficiency	*E* (MW)	100		
	η_I	33.3 %		
SF reference efficiency	$\eta_{SF,1}$	20 % (biomass)		
	$\eta_{SF,2}$	35 % (coal)		
Co-combustion MF performance data	*Cases*	*A*	*B*	*C*
	F_1 (MW)	100	50	10
	F_2 (MW)	200	250	290
	SI > 0	OK	OK	NO
	SI_1	0.30	0.25	–
	SI_2	0.40	0.36	–
	Approach #1			
	E_1 (MW)	33.3	16.7	3.3
	E_2 (MW)	66.7	83.3	96.7
	Approach #2			
	E_1 (MW)	27.3	12.2	–
	E_2 (MW)	72.7	87.8	–

MF multi-fuel, *SF* single-fuel

The obtained results show that by adopting Approach #2, the relative contribution to the total output of fuel F_1 (biomass, which is characterized by a lower SF conversion efficiency) decreases in comparison with Approach #1.

7.4 Conclusions

In this chapter, a few proposals have been made to evaluate the performance of an MF system and to quantify the contribution of each input fuel to the system output. In particular, the concept of relative SI expresses the benefit of each input fuel to the MF system by excluding the effects of all other fuels in comparison with a reference scenario. On the other hand, the global MF SI evaluates the overall performance, giving information about the synergy effect of all the input fuels. The proposed indexes allow one (i) to investigate the best way to measure the MF integrated system conversion efficiency; (ii) to measure the benefit of the integration in comparison with the SF scenario; (iii) to establish a criterion for the selection of the best configuration of integrated MF systems, for example, when various integrated plant layouts have to be compared. Moreover, two different approaches have been discussed to allocate the useful output to each input fuel. The introduced efficiency indexes have been discussed and applied to a specific co-combustion power plant case. It has been highlighted that the convenience of the co-combustion case is linked to the choice of the reference scenario. Besides co-combustion, the introduced performance assessment methodologies can also be applied in the case of topper–bottomer-integrated hybrid MF energy systems.

References

1. IEA World Energy Outlook (2010) http://www.worldenergyoutlook.org/media/weo2010.pdf. Accessed June 2013
2. Branchini L, Perez-Blanco H (2012) Handling wind variability using gas turbines. Proceedings ASME Turbo Expo. 6. pp 727–734. doi:10.1115/GT2012-68045
3. Branchini L, Perez-Blanco H (2012) Computing gas turbine fuel consumption to firm up wind power. Proceedings of ASME Turbo Expo 2012, Vol 6. pp 735–741; doi:10.1115/GT2012-68046.
4. Al-Mansoura F, Zuwala J (2010) An evaluation of biomass co-firing in Europe. Biomass Bioenergy 34(5):620–629
5. Savolainen K (2003) Co-firing of biomass in coal-fired utility boilers. Appl Energy 74(3–4):369–381
6. Bianchi M, De Pascale A (2011) Bottoming cycles for electric energy generation: parametric investigation of available and innovative solutions for the exploitation of low and medium temperature heat sources. Appl Energy 88(5):1500–1509
7. Branchini L, De Pascale A, Peretto A (2013) Systematic comparison of ORC configurations by means of comprehensive performance indexes. Appl Therm Eng 61(2):129–140. doi:10.1016/j.applthermaleng.2013.07.039
8. Bianchi M, Branchini L (2012) Advanced waste-to-energy power plant: integration with gas turbine. International Conference on Applied Energy, ICAE 2012, July 5–8 2012, Suzhou, China, Paper ID: ICAE2012–A10330
9. Campana F, Bianchi M, Branchini L, De Pascale A, Peretto A, Baresi M, Rossetti N, Fermi A, Vescovo R (2013) ORC waste heat recovery in European energy intensive industries: energy and GHG savings. Energy Convers Manag 76:244–252
10. Tillman DA (2000) Biomass cofiring: the technology, the experience, the combustion consequences. Biomass Bioenergy 19(6):365–384
11. Donatini F, Zamparelli C, Maccari A, Vignolini M (2007) High efficiency integration of thermodynamic solar plant with natural gas combined cycle. International Conference on Clean Electrical Power, ICCEP '07
12. CNIM and SENER (2005) Bizkaia waste to energy plant project, summary report, Contract No BM 80/96 ES-PO
13. Unda JI (2009) Steps forward in energy upgrading of waste to energy plants (WTE), Dubai, May 27th 2009 Middle East Waste Summit
14. Bianchi M, Branchini L, De Pascale A, Falchetti M, Fiore P (2013) Advanced waste-to-energy steam cycles. 68th Conference of the Italian Thermal Machines Engineering Association, ATI2013. To be published in Energy Procedia
15. European Commission (2006) Integrated pollution prevention and control reference document on the best available techniques for waste incineration, August 2006

Chapter 8
Specific Application Cases with GT Commercial Units

In this final chapter, detailed application cases of waste-to-energy (WTE) integrated with different commercial gas turbine (GT) units are shown and discussed. Detailed modifications of the WTE cycle and the resulting enhancement of its performance are presented. The analysis shows that combining WTE and GTs provides power output increase of up to 80 % and a steam turbine (ST) mass flow increase of up to 50 %, compared with a state-of-the-art midsize reference WTE with an efficiency of 27 %. A first law efficiency increase in the range of 8–15 percentage points can occur, depending on the GT model and layout selection. Results of waste-to-energy–gas turbine (WTE–GT)-specific application cases suggest issues and useful guidelines to (i) create new advanced WTE–GT integrated power plants or to (ii) repower existing low-performing WTE power plants to increase waste conversion into energy.

8.1 Midsize WTE Reference Steam Cycle

In order to perform a numerical investigation on the potential of a WTE–GT integrated power plant, a reference stand-alone WTE has been defined. This standard state-of-the-art plant represents the starting point for the implementation of the proposed integration with GT. The following aspects have been taken into account and fixed in order to characterize the reference WTE: (i) a given size in terms of municipal solid waste (MSW) input capacity, (ii) a given thermodynamic arrangement in terms of layout, and (iii) a standard level of steam cycle parameters of the WTE. The stand-alone WTE size and performance data are collected in Table 8.1. A schematic layout of the reference WTE steam cycle is shown in Fig. 8.1. Assumed steam cycle thermodynamic parameters are summarized in Table 8.1.

The assumed capacity, equal to 400 t/day of input MSW, is typical of a medium-sized WTE plant, and it is in line with the average values of data presented in Chap. 4 (see Fig. 4.1).

L. Branchini, *Waste-to-Energy*, DOI 10.1007/978-3-319-13608-0_8

Table 8.1 Main data of the reference stand-alone WTE plant

Reference WTE size and capacity data	
Daily capacity [t/d]	400
Annual capacity [t/y]	135,000
MSW mean LHV [MJ/kg]	11.9
Thermal capacity [MW_t]	55
Number of lines	2
Combustion system	Grate
Gross output power [MW]	15.7
Net electric power output [MW]	14.9
Net electric efficiency [%]	27.3

WTE waste-to-energy, *MSW* municipal solid waste, *LHV* lower heating value

A superheated steam cycle without reheating and with steam extractions has been considered. This adopted cycle layout arrangement, as already detailed in previous chapters, is also quite common for medium-sized WTE power plants [1].

As shown in Fig. 8.1, the superheated steam generated inside the WTE boiler, coming from the Superheater (SH) bank located in the convective section of the boiler, is sent to the ST that features two ST bleeds: One bleed (at higher pressure) is used to feed the deaerator (DEA) and to complete the primary air preheating before combustion; the second bleed (at lower pressure) is used in the feedwater heat exchanger (HE). Water cooling, in closed loop, is used for combustion grate. After passing through the grate, the hot water is used to preheat the primary and secondary inlet air streams. The economizer (ECO) is located at the end of the WTE

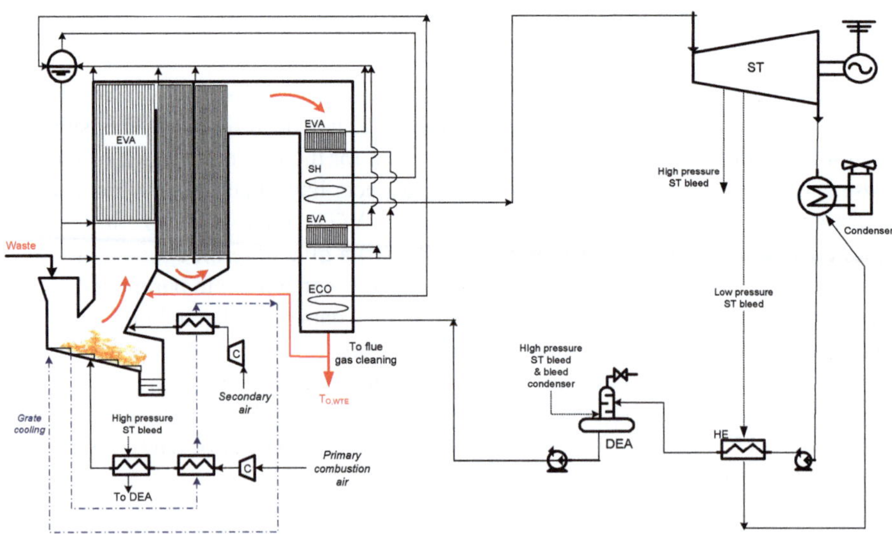

Fig. 8.1 Schematic of the reference stand-alone WTE plant. *EVA* evaporator, *SH* superheater, *ECO* economizer, *DEA* deaerator, *ST* steam turbine, *HE* general heat exchanger

Table 8.2 Steam cycle thermodynamic parameters of the reference stand-alone WTE plant

Steam turbine data	
Steam mass flow rate [kg/s]	18.9
Turbine inlet temperature [° C]	390
Turbine inlet pressure [bar]	43
Turbine isentropic efficiency [-]	0.9
ST high pressure bleed [bar]	3.5
ST low pressure bleed [bar]	1.35
Condenser pressure [bar]	0.1
WTE boiler data	
Evaporation pressure [bar]	49
Steam-superheated temperature [° C]	395
Primary air inlet temperature [° C]	120
Secondary air inlet temperature [° C]	50
Oxygen vol. dry fraction in exhaust gases [%]	7.5
FW outlet temperature [° C]	100
DEA pressure [bar]	3
Exhaust gas temperature [° C]	164
Recirculated exhaust gas rate [%]	15
Recirculated exhaust gas temperature [° C]	150
Steam cycle auxiliary consumption [MW]	0.9
Pressure losses in each heat exchanger [%]	2

boiler convective section, while the evaporator (EVA) is split in different HE banks, located in both the radiating and the convective sections of the boiler.

The steam cycle thermodynamic arrangement, the fluid properties, and the overall performance of this reference WTE plant have been fully reproduced, simulating it by means of a commercial software [2] based on a lumped parameter-modeling approach, which is used in the conventional power plant industry for complex energy system characterization and performance prediction. The software basically solves mass and energy balance equations in steady state conditions for each component (HEs, expanders, compressors, pumps, etc.), providing flows, thermodynamic states, exchanged heat and power between the system components.

Thermodynamic input for the WTE simulation, collected in Table 8.2, has been selected considering statistical data distribution of steam cycle parameters for existing WTE plants. Figure 4.1 in Chap. 4 shows superheated steam temperature and pressure of available plants, with reference to the Italian scenario. Data shown in this figure, collected in [3], have been averaged here to identify the main characteristics of the reference WTE power plant at the end of 2010 (see Table 8.1). Other input data required to simulate the WTE steam cycle, for example, steam pressure losses and heat losses in the boiler drum and distribution pipes, have been assumed in line with typical WTE steam cycle power plants.

8.2 WTE Integration with GT Units: Investigated Layout Cases and Results

The first proposed WTE steam/waterside-integration layout is shown in Fig. 8.2. The WTE–GT integrated layout is an improvement of the integrated scheme (Layout_1) presented in Chap 6 (see Fig. 6.1). A one-pressure-level (1PL) arrangement with heat recovery steam generator (HRSG) has been considered. In this layout (named *Case A*), the MSW boiler has the task of producing saturated steam, while water preheating (in the ECO) and steam superheating (in the SH) are incorporated into the HRSG. Indeed, in this proposed arrangement, in order to eliminate corrosion problems from the original SH of the WTE, the steam superheating process is moved inside the HRSG. Inside the WTE convective section, a small dryer bank replaces the original SH of Fig. 8.2. The dryer is used only to slightly increase the saturated steam temperature (of about 5 °C) to avoid steam condensation in the distribution pipes towards SH.

A preliminary parametric analysis on this integrated WTE–GT base configuration was carried out in Sect. 6.1 of Chap. 6. The additional steam mass flow rate generated in the MSW boiler caused by the integration was highlighted. This investigation in the past showed a link between the steam mass flow (m_s) and the GT size. Results of the study suggested that two key factors of the WTE–GT hybrid combined cycle (HCC) optimal design are the GT-discharged thermal power (Q_{EXH}) and the MSW input power (F_W). An optimum condition in terms of ratio of Q_{EXH} to F_W had been identified to maximize steam generation and steam-superheated temperature and to minimize exhaust gas temperature. This optimum depends on the steam cycle pressure and SH temperature.

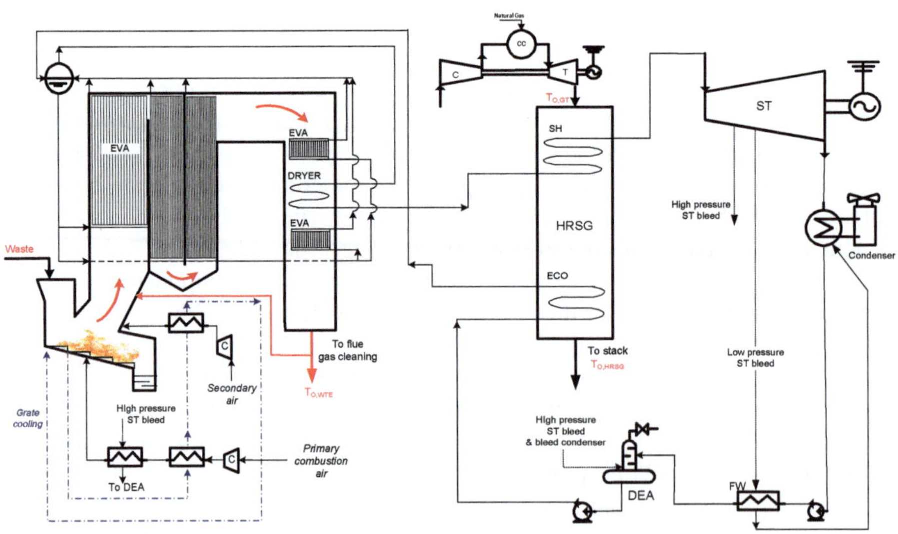

Fig. 8.2 WTE–GT integrated layout *Case A*. *EVA* evaporator, *DEA* deaerator, *SH* superheater, *HRSG* heat recovery steam generator, *ECO* economizer, *ST* steam turbine

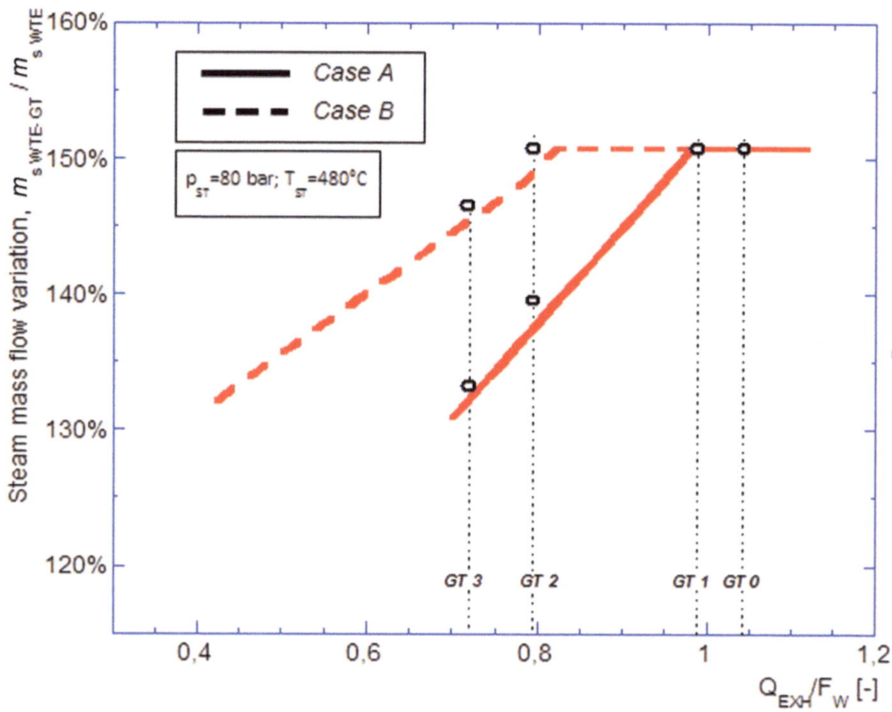

Fig. 8.3 m_s versus Q_{EXH}/F_W for three steam cycle *p–T* sets—*Case A* and *Case B* layouts; commercial gas turbine (*GT*) units are shown

Figure 8.3 shows steam production versus Q_{EXH}/F_W for the WTE–GT plant featuring the above-described *Case A* layout (continuous line). This line is obtained, according to the analysis performed in [4] and already explained in Sect. 6.1 of Chap. 6, by simulating the GT exhaust as a variable heat source and for fixed steam cycle parameters. The *p–T* values at the ST inlet are initially set equal to $p_{ST}=80$ bar, $T_{ST}=480\,°C$; these values are compatible with conventional GT exhaust temperature values (typically around 500–550 °C), considering a minimum approach of 20 °C.

A maximum value of the generated m_s can be identified; beyond this value, an increase in the HRSG thermal input (Q_{EXH}) does not lead to additional benefits, as the GT would be thermally oversized, causing a higher HRSG outlet gas temperature, that is, the GT-discharged heat is not fully exploited (see [4]). The maximum calculated m_s values of the WTE–GT plant are, in this *p–T* set case, 150 % higher than values of the steam mass flow in the reference WTE.

In order to further improve the WTE–GT integrated thermodynamic performance, an advanced configuration *(Case B),* shown in Fig. 8.4, has been also considered. Even in this second application case, the assumed integrated layout comes from an improvement of Layout_3 (see Fig. 6.20). The possibility of sharing the ECO section between WTE boiler and HRSG is proposed with a parallel arrangement. The water flow downstream DEA is split into two streams: a fraction (*f*) goes

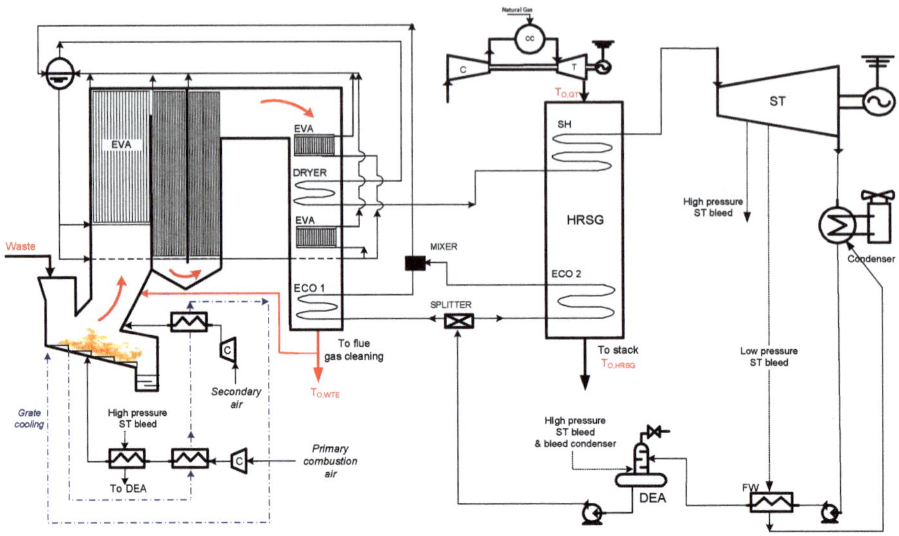

Fig. 8.4 WTE–GT integrated layout with parallel economizers—*Case B*. *EVA* evaporator, *DEA* deaerator, *ECO* economizer, *SH* superheater, *HRSG* heat recovery steam generator, *ST* steam turbine

into ECO2, that is, the ECO section inside the HRSG, while the remaining part (1-*f*) is sent to ECO1, inside the WTE boiler. A mixer is introduced before the EVA, to mix the two ECOs, outlet streams. The split factor *f* has been set in order to equalize the mixer inlets' temperatures.

The reduction in the amount of heat necessary for water economization inside the HRSG, due to the parallel ECO configuration of *Case B*, leads to a decrease in the optimal Q_{EXH}/F_W value, as shown in Fig. 8.3 (dotted line). However, the maximum generated m_s is the same for both layouts, for a given *p*–*T* set. Thus, this investigation suggests that *Case B* can be more convenient, producing the same amount of steam of *Case A*, but using a smaller GT.

8.2.1 GT Unit Selection

In order to evaluate, more accurately, the effect of specific GT machines on the WTE–GT integrated plant performance, a new numerical investigation has been performed by introducing a model-library commercial GT machine in the full WTE–GT thermodynamic model [2], with specified nominal data of the exhaust mass flow and temperature.

Different commercial units have been selected in this study, namely, GE 5251M (GT0), Rolls Royce RB211 (GT1), GE LM2500 PJ (GT2), and GE LM2500 PH (GT3). The design data of the GTs at ISO full-load condition are reported in Table 8.3 [2], while the corresponding points are shown in Fig. 8.3. The GT-discharged thermal power is also given in Table 8.3.

Table 8.3 Nameplate data of selected gas turbine (GT) units at ISO conditions

	GT0 5251M	GT1 RB211	GT2 LM2500 PJ	GT3 LM2500 PH
P_{GT} [MW]	17.8	27.1	22.6	20.8
$T_{O,GT}$ [°C]	510	506	544	533
Air flow [kg/s]	96.0	90.6	64.7	63.6
Exhaust flow [kg/s]	97.5	92.3	68.3	64.2
η_{GT} [%]	24.9	35.2	35.7	35.7
F_{NG} [MW$_t$]	71.3	77.3	63.4	58.4
Q_{EXH} [MW$_t$]	57.5	54.1	43.2	39.8
Q_{EXH}/F_W [-]	1.04	0.98	0.79	0.72

As shown in Fig. 8.3, the GT1 unit has been selected close to the optimum Q_{EXH} value (i.e., min Q_{EXH} granting max m_s) for *Case A*. In a similar way, GT2 is close to the optimum point for *Case B*. Finally, GT3 is characterized by an undersized thermal power, and GT0 is oversized for both layouts. As shown in Table 8.3, the electric efficiency of the GTs, ranging between 25 and 36%, is a key factor that strongly affects the value of Q_{EXH}. Indeed, GT0 is the smallest electric size unit, but it results in oversized Q_{EXH} because of the lowest η_{FT} value.

8.2.2 WTE–GT Integrated Plant Numerical Results

Numerical results of the WTE steam cycle integration with GT units are shown in Table 8.4, for both *Case A* and *Case B* layouts. A performance assessment has been carried out, depending on the selected GT unit and WTE–GT integrated configuration. In particular, due to the multi-fuel (MF) configuration of the investigated HCC plants, as already detailed in Chap. 7, the conversion efficiency quantification is not a simple task [5, 6]. For this reason, different key performance indicators have been taken into account to evaluate the performance of the integrated system, namely:

1. Steam cycle mass flow rate, m_s
2. HRSG effectiveness, $(T_{O,GT} - T_{O,HRSG})/(T_{O,GT} - T_{amb})$
3. ST power increase versus WTE stand-alone
4. First law efficiency, $(P_{WTE+GT})/(F_{NG} + F_W)$
5. Waste synergy index, $(P_{WTE+GT} - 0.41 \cdot F_{NG})/F_W$
6. Natural gas (NG) synergy index, $(P_{WTE+GT} - P_{WTE})/F_{NG}$
7. MF synergy index, $(P_{WTE+GT} - P_{WTE} - P_{GT})/(P_{WTE} + P_{GT})$

Among the performance indicators listed above, the first law efficiency (4) is a basic well-known index, which considers both fuels (natural gas (NG) and MSW) of equal importance. Using a different approach, the extra power, generated as a consequence of WTE–GT integration, can be assigned to the NG or MSW input

Table 8.4 Key thermodynamic variables and performance results of WTE–GTs integrated systems

	WTE stand-alone	GT0 GE 5251M		GT1 RB 211		GT2 LM2500 PJ			GT3 LM2500 PH		
		Case A	Case B	Case A	Case B	Case A	Case B	Case B-opt	Case A	Case B	Case B-opt
ST inlet temperature (T_{ST}) [°C]	390	480	480	480	480	480	480	514	480	480	490
ST inlet pressure (p_{ST}) [bar]	43	80	80	80	80	80	80	80	80	80	80
Steam-superheated temperature at SH outlet [°C]	395	485	485	485	485	485	485	519	485	485	495
Evaporation pressure at ECO outlet [bar]	49	85	85	85	85	85	85	85	85	85	85
$T_{O,GT}$ in integrated configuration [°C]	–	511	511	507	507	545	545	545	534	534	534
Approach temperature difference [°C]	–	26	26	22	22	60	60	26	49	49	39
Mass flow splitting factor (f) [–]	–	–	0.7	–	0.7	–	0.7	0.67	–	0.67	0.67
Steam mass flow rate m_s [kg/s]	18.9	28.4	28.4	28.4	28.4	26.3	28.4	27.8	25.4	27.7	27.4
Subcooling temperature difference [°C]	50	10	10	10	10	35	10	16	47	17	20
HRSG outlet temperature ($T_{O,HRSG}$) [°C]	–	162	227	139	208	138	141	138	138	138	138
HRSG effectiveness ($T_{O,GT}-T_{O,HRSG}$)/($T_{O,GT}-T_{amb}$) [–]	–	0.704	0.572	0.748	0.608	0.758	0.762	0.768	0.763	0.763	0.763
WTE outlet temperature [°C]	164	314	182	314	182	314	182	172	314	173	174
High-pressure ST bleed mass flow rate [kg/s]	1.81	2.43	2.41	2.43	2.41	2.28	2.41	2.33	2.22	2.37	2.34
Low-pressure ST bleed mass flow rate [kg/s]	1.68	2.56	2.54	2.56	2.54	2.35	2.54	2.45	2.26	2.48	2.44
Regeneration ratio	0.41	0.34	0.34	0.34	0.34	0.34	0.34	0.34	0.34	0.34	0.34
WTE–GT integrated gross output power [MW]	–	46.4	46.4	55.9	55.9	49.2	51.4	52.2	46.4	48.8	48.9
WTE–GT integrated net output power P_{WTE+GT} [MW]	–	44.9	44.9	54.2	54.2	47.7	49.7	50.6	44.9	47.2	47.3

Table 8.4 (continued)

	WTE stand-alone	GT0 GE 5251M		GT1 RB 211		GT2 LM2500 PJ			GT3 LM2500 PH		
		Case A	Case B	Case A	Case B	Case A	Case B	Case B-opt	Case A	Case B	Case B-opt
GT output power [MW]	–	17.7	17.7	27.0	27.0	22.6	22.6	22.6	20.7	20.7	20.7
ST net power [MW]	14.9	27.2	27.2	27.2	27.2	25.1	27.2	28.0	24.2	26.4	26.6
ST power increase versus WTE stand-alone [MW]	0	12.2	12.2	12.2	12.2	10.2	12.2	13.1	9.3	11.5	11.7
First law efficiency $(P_{WTE+GT})/(F_{NG}+F_W)$[%]	27.3	35.5	35.5	41.0	41.0	40.3	42.2	42.7	39.6	41.6	41.7
Natural gas synergy index $(P_{WTE+GT}-P_{WTE})/F_{NG}$ [–]	–	0.420	0.420	0.509	0.509	0.517	0.549	0.562	0.514	0.553	0.555
MF synergy index $(P_{WTE+GT}-P_{WTE}-P_{GT})/(P_{WTE}+P_{GT})$ [–]	–	0.373	0.373	0.290	0.290	0.272	0.327	0.348	0.257	0.322	0.326
Waste synergy index $(P_{WTE+GT}-0.41\cdot F_{NG})/F_W$ [–]	–	0.284	0.284	0.410	0.410	0.394	0.432	0.447	0.381	0.423	0.425
R1 $(E_{WTE}\cdot 2.6-E_{EXH}\cdot 0.21)/(0.97\cdot E_W)$ [–]	0.726	1.099	1.099	1.113	1.103	1.053	1.156	1.195	1.023	1.130	1.140

WTE waste-to-energy, *ST* steam turbine, *SH* superheater, *ECO* economizer, *HRSG* heat recovery steam generator, *WTE–GT* waste-to-energy gas turbine, *GT* gas turbine, *MF* multi-fuel (see Chap 7, Sect. 7.2.2). If the benefit is assigned to NG, the NG synergy index (6) can be used to measure the WTE–GT incremental performance, and it can be compared to the GT unit efficiency. On the other hand, the waste synergy index (5) can be used to measure the WTE–GT incremental performance compared to the WTE stand-alone. In this case, a reference constant conversion efficiency equal to 41 % of a reference 1PL combined cycle (CC) [7, 8] has been taken into account. Finally, the MF synergy index (7) can be used to measure the actual WTE–GT marginal performance in comparison with the separated reference WTE and GT power plants.

According with the investigation, the following performance results can be highlighted:

Steam mass flow rate increases significantly in all the layout configurations and GT unit cases, ranging from 34 up to 50 % of the original WTE stand-alone value (18.9 kg/s).

For fixed steam cycle *p*–*T* set (p_{ST} = 80 bar, T_{ST} = 480 °C), the points of maximum steam mass flow rate lead to maximum values of ST power increase versus WTE stand-alone. The values of extracted steam mass flows are also reported in Table 8.4 along with the calculated values of regeneration ratio.[1]

In order to additionally improve the WTE–GT integrated plant design, as the GT2 and GT3 approach temperature difference is significantly higher than typical values, a further optimization has been carried out for *Case B*, increasing the steam-superheated temperature at the SH outlet (taking into account a minimum allowed approach equal to 20 °C). In these steam-cycle-enhanced conditions (*Case B-opt*), the steam mass flow rate slightly decreases, but all the other performance indicators increase, demonstrating the possibility of optimizing the thermodynamic design of the steam cycle for GTs with a high value of $T_{O,\ GT}$. On the contrary, for both GT0 and GT1, $T_{O,\ GT}$ does not allow a further increase in the steam-superheated temperature; indeed, the approach temperature difference is already close to minimum typical values (it is equal to 26 °C and 22 °C for GT0 and GT1, respectively).

Nevertheless, all the investigated WTE–GT integrated configurations provide a gain in power output in comparison with the separate stand-alone WTE and GT systems. The ST power increase versus WTE stand-alone ranges between 9 and 12 MW, with a gain of 60–80 %. The achieved $P_{WTE\ +\ GT}$ values are 45–54 MW, that is, about three times more than the reference WTE.

The obtained values of the first law efficiency are in the range from 35 to 43 %, depending on the case, with a maximum for GT2 *Case B-opt*. In all WTE–GT cases, the first law efficiency is higher than the conversion efficiency of both the stand-alone reference WTE and the used GT unit. This is due to the synergic effect derived by integration. It follows that, for all configurations, positive values of MF synergy index are obtained.

---

[1] The regeneration ratio is calculated as the ratio of the enthalpy rise in the feedwater HEs to the enthalpy increase in the ECO section without feedwater heating.

The NG synergy index is relevant (reaching values up to about 0.55 in the best case), showing that if the benefit of integration is attributed to NG, the resulting performance index is higher than the efficiency of GT units and of reference 1PL CC (with assumed mean efficiency equal to 41 % [7, 8]).

Also, the calculated waste synergy index values are positive and *Case B* layout provides higher values for GT2 and GT3. The values of this index are higher than the WTE reference efficiency, if *Case B* with GT2 is considered.

The highest HRSG effectiveness values (and lowest HRSG outlet temperature) are achieved in *Case A*, with most of the examined GT units. On the other hand, *Case B* provides the minimum WTE exhaust temperature values, allowing for better exploitation of the MSW energy content.

Finally, as a minor result, the decrease in the GT's power output in the integrated configurations, compared to the nominal value reported in Table 8.3, is due to the presence of the HRSG's additional pressure losses, causing an increase in GT back-pressure and output power.

The application of the "R1 formula," as defined in [9], in the case of integrated WTE–GT configurations, appears questionable. The difficulty in defining a performance index for a MF energy system lies in quantifying the contribution of each input fuel to the total output energy and till date the problem is open. For the sake of completeness, results for a "modified" R1 formula have been included in Table 8.4, for each integrated configuration, assuming (i) a plant capacity factor equal to 0.9 (7800 h/year operating at full load), (ii) only the ST electrical output from the WTE section as the annual energy produced, and (iii) the electric equivalent of the GT's discharged energy recovered in the HRSG section as WTE annual "imported" energy.[2] With this adapted definition, the R1 value increases in comparison with the reference case, and it will allow all the WTE–GT integrated configurations to be classified as recovery facilities. This result suggests that the benefit of GT integration could be useful also in cases of existing, old, and low-performance WTE plants that are not satisfying the R1 threshold in the stand-alone configuration.

A final global comparison of all the analyzed cases shows that the best-performing configuration in terms of almost all the indexes is WTE-GT2 *Case B* (and in particular the *B-opt* optimized steam cycle settings), confirming that a guideline for the WTE–GT design optimization is to select the GT unit providing the minimum Q_{EXH} but, at the same time, allowing for the achievement of the maximum m_s (Fig. 8.3).

The case of an existing WTE repowered with GT will call for additional and specific evaluations; nevertheless, some general considerations can be made:

In order to accommodate the steam mass flow rate and the increase of the steam cycle parameters (temperature and pressure) increase due to the GT integration, a replacement of the original ST, condenser, and generator is necessary.

---

[2] A factor equal to 0.21 has been assumed as the electric equivalent of GT-discharged heat, in line with typical 1PL CC values [7–8].

In order to avoid original steam cycle component replacement, only an increase of up to 10% of the nominal electric power can be supported by the system working continuously in normal operating condition. One major design challenge, when repowering an existing plant, is to match the steam production capability of the topper cycle and the HRSG with the steam needs of the existing ST. Thus, the GT's optimal size, in case of a limited ST power output increase, is expected to be significantly reduced compared to the selected commercial units of Table 8.3.

As a general remark, for repowering an existing WTE plant, GT the optimal size of the GT must be chosen strictly connected to the bottomer cycle size, in order to achieve optimum plant match condition.

Despite the increased costs in comparison with the reference WTE due to ST enlargement, GT installation, and operational costs, economic benefits occur through both for the additional energy production incomes and the removed corrosion problems in the WTE convective SH section. These economic aspects have not been quantified in this study, but they are in accordance with the field data from the only existing WTE–GT plant in Europe [10–12]. The planned maintenance intervals for corrosion problems can be reduced to a third in comparison with stand-alone WTE.

Finally, the resulting cost of electricity generated by the integrated plant is mainly linked to the method used to allocate the power output to each input fuel [5]. The first law efficiency can be used to allocate the generated power output to waste and NG in a proportional way. For example, in GT2 *Case B-opt*, $P_{NG}=[P_{WTE+GT}/(F_{NG}+F_{W})]\cdot F_{NG}=27.1$ MW and $P_{W}=[P_{WTE+GT}/(F_{NG}+F_{W})]\cdot F_{W}=25.1$ MW. Thus, a fraction of the ST power output increase is allocated to NG contribution according to this method. The result of electricity allocation affects, as a consequence, the cost of electricity.

8.3 Conclusion

A detailed numerical analysis of the achievable performance by a WTE power plant that is integrated with different existing GT units has been presented. The resulting performance enhancement compared with a reference stand-alone WTE has been quantified by considering two different layout configurations of the plant, both with 1PL arrangement of the HRSG.

Combining WTE with GTs in a hybrid CC provides an increase in the ST power output of up to 80%. This is a significant increase compared with typical mid size stand-alone WTE power plants. The steam cycle mass flow rate generation is increased by up to 50%.

Numerical results of WTE repowering show that the best GT size to match maximum performance (maximum steam generation, ST power output, and minimum exhaust gas temperature) varies depending on the layout configuration and the specific selected GT unit.

This study shows that by integrating a WTE with a commercial GT, a substantial increase in the first law efficiency occurs, compared with the separated reference WTE and GT. Moreover, considerable synergic effect is shown by our numerical evaluation of a midsize conventional WTE.

Due to the high exhaust temperature values of commercial GT units, an increase in the steam cycle parameters (i.e., steam-superheated temperature and evaporation pressure) compared with the *p–T* set of the reference WTE must be pursued in the design of the WTE–GT integrated plant. Consequently, significant power output increase can be achieved.

Finally, this study provides useful guidelines and performance benchmarks for the optimal design of future advanced WTE power plants integrated with GTs.

References

1. Martin GmbH Waste-to-Energy Plant KHKW Zürich-Hagenholz. Switzerland. http://www.martingmbh.de/
2. Thermoflex software version 22 by Thermoflow inc. (2012)
3. ENEA-Federambiente (2012) Report on energy recovery from municipal solid waste in Italy, 3rd edn. (In Italian)
4. Bianchi M, Branchini L (2012) Advanced waste-to-energy power plant: integration with gas turbine. International conference applied energy, July 5–8, 2012, Suzhou, China Paper ID: ICAE2012–A10330
5. Bianchi M, Branchini L, De Pascale A (2014) Performance indexes and output allocation for multi-fuel energy systems. Energy Proc 45:32–41. doi:10.1016/j.egypro.2014.01.005
6. Bianchi M, Branchini L, De Pascale A, Falchetti M, Fiore P (2014) Advanced waste-to-energy steam cycles. Energy Proc 45:1205–1214. doi:10.1016/j.egypro.2014.01.126
7. Gas Turbine World (2004–2005) GTW handbook, vol 24. Pequot, Southport, Pequot Publication
8. Bhargava RK, Bianchi M, Campanari S, De Pascale A, di Montenegro GN, Peretto AA (2010) Parametric thermodynamic evaluation of high performance gas turbine based power cycles. J Eng Gas Turbines Power 132(2). Article no. 022001. doi:10.1115/1.3155782
9. Directive (22.11.2008) 2008/98/EC of the European Parliament and of the Council of 19 November 2008 on waste. Official Journal European Union OJ L 312/3, 22.11.2008
10. CNIM and SENER (2005) Bizkaia waste to energy plant project, summary report. Contract No. BM 80/96 ES-PO
11. SENER Ingeniería y Sistemas S.A. High efficiency technologies applied for waste to energy (WTE)
12. AZN and EPZ Combination of a waste incineration plant and a combined cycle power plant. Project no. BM/349/93/NL

Index

L. Branchini, *Waste-to-Energy*, DOI 10.1007/978-3-319-13608-0

MIX
Papier aus verantwortungsvollen Quellen
Paper from responsible sources
FSC® C105338

If you have any concerns about our products,
you can contact us on
ProductSafety@springernature.com

In case Publisher is established outside the EU,
the EU authorized representative is:
Springer Nature Customer Service Center GmbH
Europaplatz 3, 69115 Heidelberg, Germany

Printed by Libri Plureos GmbH
in Hamburg, Germany